GERMANIDES

GERMANIDY

ГЕРМАНИДЫ

GERMANIDES

Grigorii V. Samsonov

Director, Laboratory of Metallurgy of Rare Metals and Refractory Compounds
Institute of Cermets and Special Alloys
Academy of Sciences of the Ukrainian SSR
Kiev, USSR

and

Vladimir N. Bondarev

Institute of Physicochemical Principles of Mineral Raw Material Processing
Siberian Branch, Academy of Sciences of the USSR
Novosibirsk, USSR

Translated from Russian

Springer Science+Business Media, LLC 1969

Grigorii Valentinovich Samsonov was born in 1918. An internationally recognized scientist, he was elected to the Ukrainian Academy of Sciences as an associate member in 1961. Professor Samsonov has been associated with various institutes and laboratories in Moscow and Kiev, including the Moscow Institute of Fine Chemical Technology, the Moscow Institute of Nonferrous Metals and Gold, Kiev State University, and the Materials Science Institute of the Academy of Sciences of the Ukrainian SSR. He has held the posts of head of the Department of Powder Metallurgy and Rare Metals at the Kiev Polytechnic Institute and deputy director of the Institute of Scientific Studies. He is the author of more than 500 publications and inventions in the fields of inorganic and physical chemistry, crystallography, solid state physics and chemistry, metallurgy of rare metals, and powder metallurgy. His work has been translated from Russian into seven languages.

Vladimir Nikolaevich Bondarev was born in 1933. He graduated from the Moscow Institute of Steel in 1957, specializing in physicochemical investigation of metallurgical processes. From 1957 to 1961, he worked as a metallurgical engineer at the Dneprov Titanium-Magnesium Plant, and since 1961, he has been a member of the scientific staff of the Institute of Physicochemical Principles of Mineral Raw Material Processing of the Siberian Branch, Academy of Sciences of the USSR. His work is devoted mainly to methods of synthesizing different types of intermetallic and high-melting compounds, and to studies of their properties.

The original Russian text, published by Metallurgiya Press in Moscow in 1968, has been corrected by the authors for the present edition.

Самсонов Григорий Валентинович
Бондарев Владимир Николаевич

ГЕРМАНИДЫ

Library of Congress Catalog Card Number 70-79918

ISBN 978-1-4899-4823-6   ISBN 978-1-4899-4821-2 (eBook)
DOI 10.1007/978-1-4899-4821-2

FOREWORD

Many rapidly developing branches of modern technology require new materials having a complicated set of physical and chemical properties, particularly dielectric or semiconductive properties, high chemical stability, and high or low reactivity under certain conditions. The use of various high-melting compounds of metals with metals, metals with nonmetals, and metals with semimetals, which can operate under high-temperature conditions while retaining the requisite physical and electrical properties, is of great interest in this connection.

In addition to the relatively well-studied high-temperature compounds of carbon and silicon (carbides and silicides), the next group of compounds, that of germanium with various elements of the periodic system (germanides) is extremely interesting. Germanides have lower melting points then carbides or silicides, but many of them are distinguished by electrophysical or magnetic properties of importance for modern engineering. The data on germanium compounds are comparatively sparse and are scattered over numerous soures, which makes it impossible to get an overall view of this important class of inorganic compounds their properties, and their possible areas of application. The few special surveys in the literature on germanium, its technology, and its properties do not fill this gap since in the main they offer only descriptions of the chemical properties and reactivity of germanium [1-42].

This book attempts a more complete review of the information on germanium compounds, the basic theories of their electronic structure, chemical bonding, and crystal chemistry, methods for producing them, and their possible areas of application. Since the term "germanides" properly refers to compounds of germanium with more electropositive elements (principally metals), the use of this term for germanium compounds with more electronegative elements seems inappropriate. Moreover, since germanium does not form compounds with a number of these elements, the authors deemed it more correct to call the section dealing with nonmetals and semimetals "Compounds of Germanium with ... Elements." This made it possible to maintain a uniform style throughout the book and to title it "Germanides," which indicates its main topic, i.e., the combination of germanium with metals to form corresponding metalloid compounds.

The paucity of research on the physical and chemical properties of germanium compounds permits us to give only a preliminary notion of the nature of these properties.

The compounds of germanium with halogens and hydrogen have not been described in separate chapters, since their properties differ rather markedly from those of other germanium compounds. Moreover, an extensive special literature has been devoted to germanium halides and it would be inexpedient to duplicate it. This first attempt to generalize the data on germanides cannot help but have deficiencies in the presentation and interpretation of the factual material and in the methodological area. The authors would be grateful for any criticisms or recommendations.

v

In conclusion, the authors wish to express their gratitude to the Soviet and foreign specialists who provided them with copies of original and often very rare works on germanides; these have made it possible to give sufficiently complete descriptions of the individual compounds.

CONTENTS

CONTENTS

CHAPTER 1

STRUCTURAL AND PHYSICAL PROPERTIES
OF GERMANIDES

STRUCTURAL CHARACTERISTICS AND CRYSTAL CHEMISTRY

Germanium occupies an intermediate position between silicon and tin, its analogs, in the periodic table. This position is manifested in many of the properties of germanium, the best known of which are its semiconductive properties. Germanium is similar in structure to diamond. Calculations [43, 44] have shown that it more closely resembles silicon than gray tin. It is known from structural chemistry that the atomic radius of germanium is closer to that of silicon that to that of gray tin. The atomic radii of some elements are given below, in nm*:

Al	0.143	Si	0.134	F	0.13
Ga	0.139	Ge	0.139	As	0.148
In	0.157	Sn	0.158	Sb	0.161

Although they are not precise physical (invariant) quantities, the atomic radii of the elements are nevertheless, together with the geometric representation of atomic symmetry and positioning, the most important parameter in crystal chemistry. According to Pauling [45], the simple-bond radii for germanium and silicon differ substantially less than those for germanium and gray tin. This becomes even clearer if we compare the radii of the tetravalent anions, as in compounds of the $Mg^{2+}X^{4-}$ class (where x is Si, Ge, or Sn):

Si^{4-}	0.271
Ge^{4-}	0.272
Sn^{4-}	0.294

There is also a similarity between compounds of germanium with metals, i.e., germanides, and the corresponding compounds of silicon, i.e., silicides. Transition-metal germanides are similar to silicides in structure and a number of physicochemical properties: they are capable of entering the superconductive state, have a rather high electrical conductivity, metallic luster, and characteristic fracture, etc.

However, germanides and silicides, in contrast to carbides and nitrides, cannot be classified as compounds of the interstitial-phase type, a term that refers to phases formed by introduction of nonmetal atoms with small atomic radii (satisfying the Hegg condition $r_x: r_{Me} < 0.59$) into the octahedral or tetrahedral pores of the hexagonal and cubic lattices of metals [46].

---

* Lattice constants and atomic radii are given in nanometers (nm) in the International system of Units: 1 nm = 10 Å.

The ratio of the radius of the germanium atom (0.139 nm) to the radii of transition-metal atoms substantially exceeds the critical Hegg ratio of 0.59 and is greater than unity in a number of cases:

Scandium	0.84	Rhodium	1.03
Titanium	0.95	Palladium	1.01
Vanadium	1.0	Lanthanum	0.74
Chromium	1.09	Lutecium	0.79
Manganese	1.07	Hafnium	0.87
Iron	1.1	Tantalum	0.95
Cobalt	1.11	Tungsten	0.99
Nickel	1.12	Rhenium	1.01
Yttrium	0.76	Osmium	1.02
Zirconium	0.86	Iridium	1.02
Niobium	0.95	Platinum	1.0
Molybdenum	1.0	Thorium	0.77
Ruthenium	1.03	Uranium	0.90

This results in replacement of metal atoms by germanium atoms rather than interstitial introduction of the latter.

In accordance with the classical rules of isomorphism, substitution leads to formation of metallic structures at ratios $r_{Ge} : r_{Me} \geq 0.95-0.96$ and germanium contents of the order of 20-32 at-%. At higher germanium contents and for metals having larger atomic radii (scandium, yttrium, titanium, zirconium, hafnium, niobium, tantalum, uranium, lanthanides, and actinides), replacement of metal atoms by germanium leads to substantial changes in the type of lattice and formation of complex structures in which covalent bonding between the germanium atoms plays a considerable role. The tendency toward formation of complex structures increases as the ratio $r_{Ge} : r_{Me}$ decreases. For metal germanides with a ratio $r_{Ge} : r_{Me} \geq 0.95-0.96$, an increase in this ratio causes simplification of the structural elements made up of germanium atoms. The most complex framework configuration of germanium atoms occurs in the germanides of lanthanides ($r_{Ge} : r_{Me} = 0.74-0.79$), thorium (0.77), uranium (0.90), and plutonium; simpler configurations (laminar structural elements composed of germanium atoms) occur in the germanides of molybdenum, niobium, and tantalum; chain-like elements occur in the germanides of titanium, zirconium, and calcium, while elements consisting of isolated pairs of germanium atoms occur in Th_3Ge_2.

Germanides with Metallic Structures. Germanides of this structural type are formed when germanium atoms replace metal atoms and the condition $r_{Ge} : r_{Me} \geq 0.95-0.96$ is satisfied. All these germanides (Cr_3Ge, Mo_3Ge, Nb_3Ge, and B_3Ge) have a structure of the β-W type, i.e., with dense cubic packing (Fig. 1).

Each metal atom is in the center of an irregular tetrahedron of germanium atoms, while each germanium atom is surrounded by 12 metal atoms forming an icosahedron. The great compactness of the lattice emphasizes the metallic character of the Me-Ge bond in germanides with a structure of the β-W type. However, Millner [47] has pointed out that there are covalent bonds as well as metallic bonds between the tungsten atoms in the β-W structure. A necessary condition for formation of an A_3X phase with a β-W structure is that the A atoms be capable of furnishing a minimum of two electrons for organization of covalent bonds, in addition to the electrons provided for the metallic bond.

Germanides with Complex Structures. Germanium, like silicon, exhibits a tendency toward formation of framework lattices capable of producing isolated groups of atoms, which

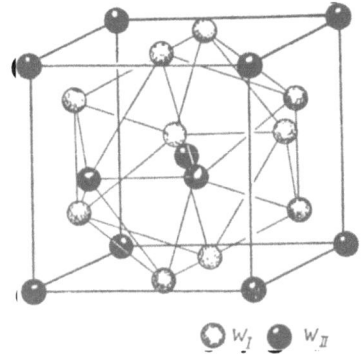

$\bigcirc W_I$ $\bullet W_{II}$

Fig. 1. Structure of β-W type [50, 51].

are incorporated as separate structural elements into the lattices of certain germanides.

In the same manner as silicides, germanides can be classified into the following structural types [50, 51]:

a) with isolated germanium atoms;

b) with isolated pairs of germanium atoms;

c) with extremely dense layers of atoms of both components;

d) with chains of germanium atoms;

e) with layers of germanium atoms;

f) with a framework of germanium atoms.

Structures with Isolated Germanium Atoms

The FeSi structure is a cubic cell with four iron atoms and four silicon atoms. It produces heptagons of silicon atoms with iron atoms in their centers. The chemical compound CrGe is of this structural type. According to Esslinger and Schubert [52], the compounds $CoGe_{0.75}Ga_{0.25}$, $RhGe_{0.80}Ga_{0.20}$, $NiGe_{0.16}Ga_{0.84}$, and $PdGe_{0.20}Al_{0.80}$ are also of this type.

Structures with Isolated Pairs of Germanium Atoms

In studying the germanides of thorium and uranium, a great similarity was observed between these compounds and the corresponding silicides. The compounds UGe_3 and Th_3Ge_2 are isostructural with USi_3 and Th_3Si_2.

The U_3Si_2 structure is a tetragonal cell with six uranium atoms and four silicon atoms. The pairs of silicon atoms are parallel to the (001) plane, while the uranium atoms form layers alternating in the sequence ...AA... to produce trigonal and tetragonal prisms. Uranium atoms occupy the centers of the tetragonal prisms, while pairs of silicon atoms occupy those of the trigonal prisms. The compound Th_3Ge_2 crystallizes with this structure. Pairs of germanium atoms separated by a distance of 0.248 nm are present in the thorium lattice in this compound. Pair formation is thus not as pronounced as for the corresponding silicides. As a result, the Ge−Ge bond in the crystal lattice is not as strong as the Si−Si bond [48].

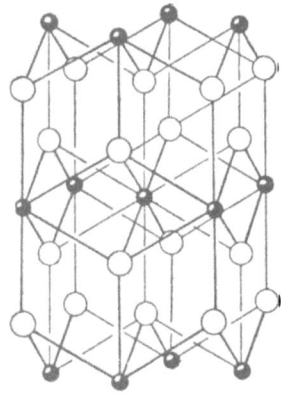

$\bullet Mo$ $\bigcirc Si$

Fig. 2. Structure of $MoSi_2$ type [50, 51].

Structures with Dense Layers of Atoms of Both Components

The $MoSi_2$ structure is a tetragonal cell with two $MoSi_2$ molecules (Fig. 2). The positions of the atoms are

Mo 2:(a);

Si 4:(c), $z = 1/3$.

The silicon atoms form a framework, whose voids contain molybdenum atoms. Twin densely-packed layers of silicon alternate with layers of molybdenum atoms along the Z axis. This structure can also be regarded as consisting of layers parallel to the (010) plane and having extremely dense hexagonal packing. The layers alternate in the sequence ...ABAB..., with the B layer displaced by 0.5a along the X axis. The compound $MoGe_2$ has a structure of this type.

The $CrSi_2$ structure is a hexagonal cell with three formal units of $CrSi_2$. The positions of the atoms are

$$Cr\ 3:(d);$$

$$Si\ 6:(j),\ x = 1/6.$$

The germanides $NbGe_2$ and $TaGe_2$ crystallize with a structure of this type.

The $TiSi_2$ structure is a rhombic cell with eight $TiSi_2$ molecules (Fig. 3). The positions of the atoms are

$$Ti\ 8:(a);$$

$$Si\ 16:(c),\ x = 1/3.$$

The silicon and titanium atoms form layers with dense packing; each titanium atom in the layer is surrounded by six silicon atoms, while each silicon atom is surrounded by three titanium atoms. The layers are arranged one above the other in such fashion that the center of a titanium atom in the second layer is over point 1, that of an atom in the third layer is over point 2, and that of an atom in the fourth layer is over point 3. The compound $TiGe_2$ crystallizes with a structure of this type.

Structures with Chains of Germanium Atoms

The Mn_5Si_3 structure is a hexagonal cell with two Mn_5Si_3 molecules. The positions of the atoms are

$$Mn_I\ 4:(d);$$

$$Mn_{II}\ 6:(g),\ x_{Mn} = 0.23;$$

$$Si\ 6:(g),\ x_{Si} = 0.06$$

Six of the ten manganese atoms form a distorted octahedron with the silicon around the vacant points (000) and $(00\frac{1}{2})$. The germanides Mn_5Ge_3, Ti_5Ge_3, U_5Ge_3, and Zr_5Ge_3 have a structure of this type.

Structures with Layers of Germanium Atoms

The $ZrSi_2$ structure is a rhombic cell with four formal units of $ZrSi_2$. The positions of the atoms are

$$Zr\ 4:(c),\ y_{Zr} = 0.106;$$

$$Si_I\ 4:(c),\ y_I = 0.750;$$

$$Si_{II}\ 4:(c),\ y_{II} = 0.355.$$

The silicon atoms form chains running parallel to the X and Z axes. The compounds $ZrGe_2$ and UGe_2 are of this type. However, the structural elements of $ZrGe_2$ are less zigzag in character than those of $ZrSi_2$, since the $r_{Zr}:r_{Ge}$ ratio is lower.

The $CaSi_2$ structure is a rhombic cell with two formal units of $CaSi_2$. The atoms are

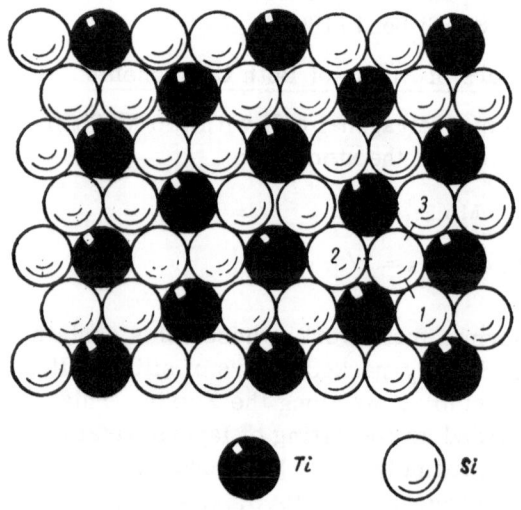

Fig. 3. Structure of $TiSi_2$ type [50, 51].

$$Ca\ 2:(c),\ x_{Ca} = 0.083;$$

$$Si_I\ 2:(c),\ x_I = 0.185;$$

$$Si_{II}\ 2:(c),\ x_{II} = 0.352;$$

Corrugated layers of silicon atoms alternate with layers of calcium atoms. The layers of silicon atoms alternate in the sequence ...ABCACABCBC... AB..., producing dense twelve-layer packing. The compound $CaGe_2$ crystallizes in this manner. A characteristic radius ratio $r_{Me}:r_{Ge}$ of about 1.5 has been found for this structure. Germanium has a negative charge with respect to the positive metal ions in such lattices, which is in good agreement with the observed Ge−Ge distances [53].

Structures with a Framework of Germanium Atoms

The $ThSi_2$ structure (Fig. 4) is a tetragonal cell with four formal units of $ThSi_2$. The silicon atoms form zigzag chains passing through prisms of thorium atoms parallel to the X and Y axes at different heights. The projections of the chains are directed toward one another; the distance between the silicon atoms in the projections equals that between the silicon atoms in the chain, i.e., a three-dimensional framework of silicon atoms is created. The compounds $PrGe_2$, $PuGe_2$, and α-$ThGe_2$ crystallize in this manner. The ability of germanium to form chains is somewhat hampered in β-$ThGe_2$, since its composition does not precisely correspond to the formula AX_2.

The structure of the germanides $FeGe_2$ and $CoGe_2$ differs from that of the corresponding silicides. The compound $FeGe_2$ is of the $CuAl_2$ type (Fig. 5), while $CoGe_2$ [53] is of a type intermediate between the latter compound and feldspar. To some extent, these germanides of Group 8a exihibit structural types in which the valence of such compounds as $Me^{8+}Ge_2^{4-}$ is retained, although such ions do not exist.

It has been hypothesized that the charge on these ions is due to polarization of the electrons [54]. The radius ratio $r_{Me}:r_{Ge}$ ranges from 0.75 to 0.85 for this structure. Polarity leading to formation of germanium anions is more than probable.

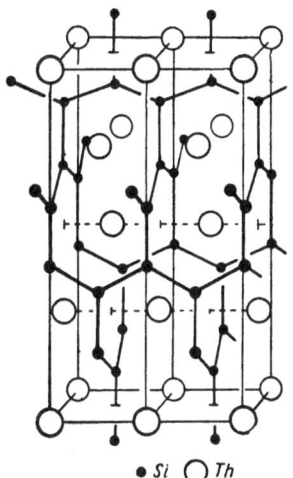

• Si ◯ Th

Fig. 4. Structure of
$ThSi_2$ type [50, 51].

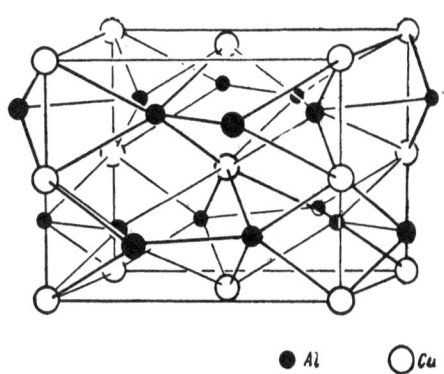

● Al ◯ Cu

Fig. 5. Structure of $CuAl_2$ type
[50, 51].

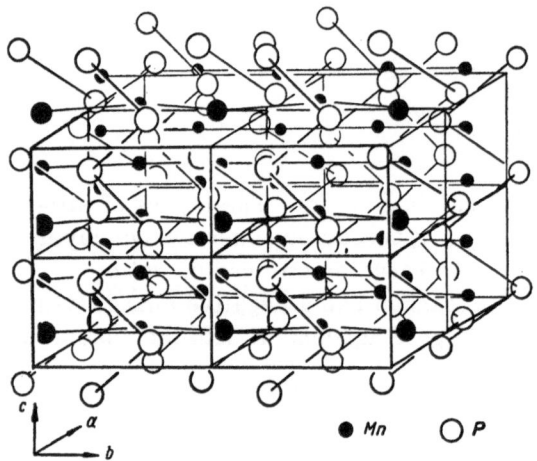

Fig. 6. Structure of MnP type [50, 51].

Among the germanides with a high content of an electropositive metal, we must mention Mg_2Ge. This germanide is isostructural with the compounds Mg_2Si and Mg_2Sn. Ternary compounds, such as Li_5GeP_3 and Li_5GeAs_3, which has a feldspar structure, are also included in this class. They are similar to Mg_2Ge, but the germanium must be regarded as a cation $(Li^{5+}Ge^{4+})P_3^{3-}$.

Like Ca_2Si, the compound Ca_2Ge crystallizes with a structure of the $PbCl_2$ type [55]. Each germanium atom is surrounded by eight calcium atoms, which form polyhedra with 14 faces. There is a large decrease (by about 9%) in the radius of the calcium in this case. On the other hand, the radius of the germanium (0.124 nm) is also reduced. This is not at all obligatory for the anionic germanium in these compounds.

Other Types of Germanides. The NiAs structure is a hexagonal cell with two formal units of NiAs. The positions of the atoms are

$$Ni \; 2:(a);$$

$$As \; 2:(c).$$

The arsenic atoms form a densely-packed hexagonal lattice, all the octohedral voids of which are occupied by nickel atoms. The compounds Co_2Ge, Fe_2Ge, Mn_2Ge, and Ni_2Ge crystallize in this manner.

The MnP structure is a rhombic cell with four formal units of MnP (Fig. 6). The positions of the atoms are

$$Mn \; 4:(c), \; xy\,^1/_4; \; ^1/_2 - x, \; ^1/_2 + y, \; ^1/_4; \; xy\,^3/_4; \; ^1/_2 + x, \; ^1/_2 - y, \; ^3/_4; \; x_{Mn} = 0.20, \; y_{Mn} = 0.005;$$

$$P \; 4:(c), \; x_P = 0.57, \; y_P = 0.19.$$

The MnP structure has a rhombic distortion of the NiAs type. The B 20 structure with a coordination number of 7 is intermediate between monogermanides containing Ge–Ge chains (with a coordination number of 6) and Me–Me chains (with a coordination number of 6). The

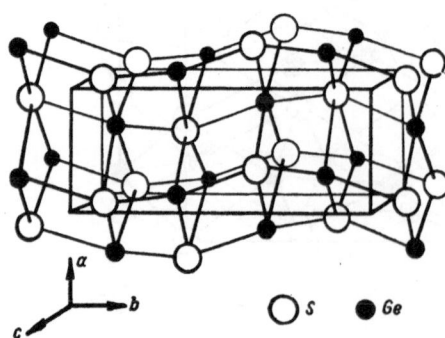

Fig. 7. Structure of GeS type [50, 51].

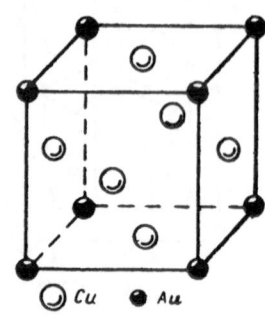

Fig. 8. Structure of Cu_3Au type [50, 51].

compounds RhGe, IrGe, NiGe, PdGe, and PtGe have an MnP structure, a variety of the NiAs lattice. Zhdanov and Glagoleva [56] report an altered lattice of the MnP type. The germanides listed above correspond to the appropriate silicides.

The GeS structure (Fig. 7) is a rhombic cell with four formal units of GeS. The positions of the atoms are

$$\text{Ge } 4\text{:(c)}, \; x\,y\,\tfrac{1}{4}; \; \tfrac{1}{2} - x, \; \tfrac{1}{2} + y, \; \tfrac{1}{4}; \; x\,y\,\tfrac{3}{4}; \; \tfrac{1}{2} + x, \; \tfrac{1}{2} - y, \; \tfrac{3}{4}; \; x_{\text{Ge}} = 0.167, \; y_{\text{Ge}} = 0.375;$$

$$\text{S } 4\text{:(c)}, \; x_{\text{S}} = 0.111, \; y_{\text{S}} = 0.139.$$

This structure is a severely distorted version of the NaCl lattice.

The compound ThGe, like Th(O, N, C, S, P), U(O, N, S, C, P, As), and other actinide monocompounds, has a structure of the NaCl type and not a monoboride structure, as USi does. Formation of Ge−Ge or Si−Si chains is hampered by the presence of thorium, whose atoms occupy a larger volume than uranium atoms. The Ge^{4-} ion should not exist even when the germanium atoms in ThGe are in contact, since the Ge−Ge distance of 0.427 nm is too short. The compound ThGe decomposes peritectically into Th_3Ge_2 and β-$ThGe_2$. It has a structure intermediate between a purely heteropolar structure and one of the Hegg interstitial-phase type. This is confirmed by the metallic character of the compound.

The Cu_3Au structure (Fig. 8) is a cubic cell with one formal unit of Cu_3Au. The positions of the atoms are

$$\text{Au } 1\text{:(a)};$$

$$\text{Cu } 3\text{:(c)}.$$

The gold atoms are located at the vertices of the cell, while the copper atoms are in the centers of the faces. The gold atoms lie in the centers of cubic octahedra of copper atoms. The structure is composed of these octahedra and an equal number of vacant octahedra. The cubic octahedra lie precisely one above the other along the fourth-order axis. The compounds Ni_3Ge, $PuGe_3$, and UGe_3 crystallize in this manner.

The structure of $ThGe_3$ is still unknown but has a pseudocubic cell with a = 1.172 nm [48].

Compounds corresponding to the formula A_5X_3 form three types of structures: Mo_5Si_3 (T1), $Mo_5(Si, B)_3$ (T2), and (Mo, Si)$_5$Si, C)$_3$ (D8$_8$).

The compounds Ti_5Ge_3, Zr_5Ge_3, $Nb_5Ge_3(C)$, $Ta_5Ge_3(C)$, Mn_5Ge_3, and V_5Ge_3 crystallize in a structure of the D8$_8$ type. The germanides of tantalum and niobium are stabilized by carbon. Metalloids probably also exert a stabilizing influence in titanium and zirconium phases, but the amounts required are very small in these cases. The amount of stabilizer (C, N, O, or B) in germanides is naturally larger than in silicides [48].

The compounds Nb_5Ge_3, β-Ta_5Ge_3, Cr_5Ge_3, and Mo_5Ge_3 crystallize with a structure of the Mo_5Si_3 (T1) type, while α-Ta_5Ge_3 has a structure of the $Mo_5(Si, B)_3$ type. This ternary phase, like W5(Si, B)$_3$, was discovered by Nowotny et al. [57]. Parthe et al. [58] reported on the geometry of these crystal lattices and their relationship to the C_{16} (CuAl$_2$) structure and to structures of the feldspar type.

It follows from analysis of the interatomic distances that phases with a D8$_8$ structure should have a more polar character than those with a T1 or T2 structure (provided that the germanium forms anions).

The class of A_5X_3 compounds extends to Fe_5Si_3 in the first long period for silicides, but only to Mn_5Ge_3 for germanides. According to Millner [47], the superconductor Rh_5Ge_3 has a different structure. However, there is a great similarity between Rh_5Ge_3 and Rh_2Ge, which crystallizes in an ideal structure of the NiAs type.

TABLE 1. Crystal Structure of Germanides

Metal	Germanide	Germanium content, % by mass	Structure	Structural type	Space group	Lattice constants, nm — a	b	c	c/a	Density, g/cm³ — X-ray	Pycnometric	Number of formal units	References
Li	$Li_{22}Ge_5$	70.39	Cubic	$Li_{22}Pb_5$	T^a	1.886	—	—	—	—	—	—	[152]
	$Li_{15}Ge_4$	73.61	»	Cr_3Si	T_d^6	1.0783	—	—	—	—	—	—	[60, 151]
	Li_9Ge_2	77.71	—	—	—	—	—	—	—	—	—	—	[59]
	LiGe	91.27	Tetragonal	—	—	0.438	—	0.580	1.32	—	—	—	[91]
Na	NaGe	75.95	Monoclinic	—	C_{2h}^5	1.233	0.670	1.142	$[\beta=2.1\ \text{rad}\ (\beta=120°)]$	3.14	3.09	—	[64]
K	KGe	64.99	Cubic	—	—	1.233	0.670	1.142	$[\beta=2.09\ \text{rad}\ (\beta=119.9°)]$	2.85	2.78	—	[157]
	KGe_4	88.13	»	—	—	1.278	—	—	—	—	—	—	[64]
Rb	RbGe	45.93	»	—	—	1.390	—	—	—	3.66	3.63	—	[64]
	$RbGe_4$	77.25	»	—	—	1.319	—	—	—	—	—	—	[64]
Cs	CsGe	35.33	»	—	—	1.400	—	—	—	—	—	—	[64]
	$CsGe_4$	68.60	»	—	—	1.367	—	—	—	4.28	4.28	2	[64]
Cu	Cu_5Ge	18.59	Hexagonal	Mg	D_{6h}^4	0.2515; 0.2660	—	0.4119; 0.4287	1.638–1.612	—	—	—	[49], [97, 98]
	$\varepsilon\text{-}Cu_3Ge$	27.58	—	—	—	0.4168	—	0.5001	1.20	—	—	—	[97, 98]
	$\varepsilon_1\text{-}Cu_3Ge$		Distorted rhombic	Ag_3Sn	D_{6h}^4	0.2645	0.4549	0.4198	—	17.03	14.00	2	[97, 98]
	$\varepsilon_1\text{-}Cu_3Ge$		Distorted monoclinic	—	—	0.2631	0.4200	0.4568	$[\beta=1.56\ \text{rad}\ (\beta=89.7°)]$	—	—	—	[160]
Mg	Mg_2Ge	59.88	Cubic	CaF_2	O_h^5	0.6386	—	—	—	3.091	3.086	4	[100]
Ca	$Ca_{33}Ge$	5.20	»	CaF_2	O_h^5	0.6380	—	—	—	—	3.082	4	[99]
	Ca_7Ge	20.56	»	—	O_h^5	1.019	—	—	—	—	2.196	1	[103]
	Ca_2Ge	47.53	»	—	O_h^7	0.945	—	—	—	—	2.825	4	[103]
	CaGe	64.43	Rhombic	$PbCl_2$	D_{2h}^{16}	0.9069	0.7734	0.4834	—	—	—	4	[55]
	$CaGe_2$	78.36	Rhombic	CrB	D_{2h}^{17}	0.4565	1.0836	0.4004	—	—	—	4	[104]
			Rhombic	CaSi	D_{1h}^{17}	0.4001	0.4575	1.084	—	—	—	4	[55]
			Rhombohedral	$CaSi_2$	D_{3d}^5	1.049	—	—	$[0.37\ \text{rad}\ (\alpha=21.7°)]$	—	—	2	[82]
Ba	BaGe	34.58	Rhombic	CrB	D_{2h}^{17}	0.507	1.198	0.430	—	—	—	4	[438, 478]
	$BaGe_3$	51.38	Cubic	—	O^7	1.452	—	—	—	—	4.28	28	[71]
Sr	SrGe	45.31	Rhombic	CrB	D_{2h}^{II}	0.482	1.139	0.4167	—	—	—	—	[478]
Sc	Sc_5Ge_3	49.20	Hexagonal	Mn_5Si_3	D_{6h}^3	0.7939	—	0.5883	0.741	2.76	—	2	[143]
	ScGe	61.75	Rhombic	CrB	D_{2h}^{17}	0.4007	1.006	0.3762	—	—	—	4	[199, 200]
	$ScGe_2$	76.36	»	$ZrSi_2$	D_{2h}^{17}	0.3868	1.4873	0.3793	—	—	—	4	[198]
Y	Y_5Ge_3	32.88	Hexagonal	Mn_5Si_3	D_{6h}^3	0.8471	—	0.6350	0.749	5.57	5.40	2	[141]

TABLE 1. (Continued)

Metal	Germanide	Germanium content, % by mass	Structure	Structural type	Space group	a	b	c	c/a	Density, g/cm³ X-ray	Density, g/cm³ Pycnometric	Number of formal units	References
Y	Y_5Ge_4	39.51	Rhombic	Sn_5Ge_4	$Pn2_1a$	0.763	1.468	0.768	—	—	—	—	[441]
	YGe	44.95	»	CrB	D_{2h}^{17}	0.4262	1.0694	0.3941	—	—	—	4	[199, 200]
	Y_2Ge_3	55.05	Hexagonal	AlB_2	D_{6h}^{1}	0.3960	—	0.4140	1.045	—	—	1	[145]
	YGe_2	62.02	»	AlB_2	D_{6h}^{1}	0.3935	—	0.4139	1.052	—	—	1	[198]
	YGe_2	—	Tetragonal	α-$ThSi_2$	D_{4h}^{19}	0.4060	—	1.3683	3.397	—	—	4	[198]
La	La_5Ge_3	23.87	Hexagonal	Mn_5Si_3	D_{6h}^{3}	0.8958	—	0.6795	0.759	3.72	—	2	[143, 202]
	La_4Ge_3	—	Cubic	Th_3P_4	—	0.9356	—	—	—	—	—	—	[440]
	LaGe	34.32	Rhombic	FeB	D_{2h}^{16}	0.845	0.413	0.610	—	—	—	4	[147, 203]
	LaGe	—	Rhombic	FeB	D_{2h}^{13}	0.8486	0.4134	0.6119	—	—	—	4	[200]
	$LaGe_2$	51.10	Tetragonal	α-$ThSi_2$	D_{4h}^{19}	0.4321	—	1.4209	3.288	—	7.059	4	[146, 368]
	$LaGe_2$	—	Rhombic	α-$GdSi_2$	D_{2h}^{28}	0.441	0.430	1.4190	—	—	—	4	[145, 146]
Ce	Ce_5Ge_3	23.71	Hexagonal	Mn_5Si_3	D_{6h}^{3}	0.8875	—	0.6570	0.740	3.92	—	2	[143, 202]
	CeGe	34.13	Rhombic	FeB	D_{2h}^{16}	0.835	0.408	0.602	—	—	—	4	[200, 203]
	CeGe	—	»	FeB	D_{2h}^{16}	0.8354	0.4082	0.6033	—	—	—	4	[434]
	CeGe	—	»	FeB	D_{2h}^{16}	0.8356	0.4084	0.6041	—	—	—	4	[435]
	$CeGe_2$	50.89	Tetragonal	α-$ThSi_2$	D_{4h}^{19}	0.4280	—	1.4037	3.279	—	—	4	[145]
	$CeGe_2$	—	Rhombic	α-$GdSi_2$	D_{2h}^{28}	0.4270	0.426	1.408	3.297	—	—	4	[146]
Pr	Pr_5Ge_3	23.61	Hexagonal	Mn_5Si_3	C_{6v}^{3}	0.880	—	0.660	0.750	—	—	2	[202]
	Pr_4Ge_3	—	Cubic	Th_3P_4	—	0.9153	—	—	—	—	—	—	[440]
	PrGe	34.0	Rhombic	CrB	D_{2h}^{17}	1.4474	1.1098	0.4064	—	7.93	6.78	4	[170, 196, 203]
	PrGe	—	»	FeB	D_{2h}^{16}	0.8288	0.4050	0.5987	—	—	—	4	[434]
	PrGe	—	»	CrB	D_{2h}^{17}	0.4479	1.1084	0.4050	—	—	—	4	[434]
	$PrGe_2$	50.75	Tetragonal	α-$ThSi_2$	D_{4h}^{19}	0.4253	—	1.3940	3.277	7.48	7.24	4	[170, 196]
	$PrGe_2$	—	»	α-$ThSi_2$	D_{4h}^{19}	0.426	—	1.398	3.281	—	—	4	[145, 146]
	Nd_5Ge_3	23.19	Hexagonal	Mn_5Si_3	D_{6h}^{3}	0.876	—	1.657	0.750	—	—	2	[202]

TABLE 1. (Continued)

Metal	Germanide	Germanium content, % by mass	Crystal-chemical properties — Structure	Structural type	Space group	Lattice constants, nm — a	b	c	c/a	Density, g/cm³ — X-ray	Pycnometric	Number of formal units	References
Nd	Nd_5Ge_4	28.70	Rhombic	Sm_5Ge_4	$Pn2_1a$	0.786	1.506	0.793	—	—	—	—	[441]
	NdGe	33.48	"	CrB	D_{2h}^{17}	0.445	1.102	0.403	—	—	—	4	[203]
			"	CrB	D_{2h}^{17}	0.4442	1.1012	0.4030	—	—	—	4	[434]
			"	CrB	D_{2h}^{17}	0.4456	1.1027	0.4035	—	—	—	4	[435]
	$NdGe_2$	50.16	Tetragonal	α-$ThSi_2$	D_{4h}^{19}	0.424	—	1.390	3.278	—	—	4	[145, 146]
			"	α-$ThSi_2$	D_{4h}^{19}	0.4258	—	1.3870	3.260	—	—	4	[192]
Sm	Sm_5Ge_3	22.46	Hexagonal	Mn_5Si_3	D_{6h}^{3}	0.865	—	0.649	0.750	—	—	2	[202]
	Sm_5Ge_4	27.86	Rhombic	Sm_5Ge_4	$Pn2_1a$	0.775	1.494	0.784	—	—	7.62	4	[441, 482]
	SmGe	32.56	"	CrB	D_{2h}^{17}	0.437	1.088	0.400	—	—	—	4	[203]
			"	CrB	D_{2h}^{17}	0.436	1.086	0.401	—	—	—	4	[434]
			"	CrB	D_{2h}^{17}	0.4387	1.0890	0.3993	—	—	—	4	[435]
	$SmGe_2$	49.12	Tetragonal	α-$ThSi_2$	D_{4h}^{19}	0.4193	—	1.3835	3.299	—	—	4	[145, 146]
Eu	EuGe	32.33	Rhombic	CrB	D_{2h}^{17}	0.472	1.121	0.412	—	—	—	4	[203]
			"	CrB	D_{2h}^{17}	0.4730	1.1194	0.4105	—	—	—	4	[435]
	$EuGe_2$	48.86	Hexagonal	—	—	0.4102	—	0.4995	1.218	6.44	6.67	—	[145, 146, 193]
Gd	Gd_5Ge_3	21.69	Hexagonal	Mn_5Si_3	D_{6h}^{3}	0.8546	—	0.6410	0.750	—	—	2	[142]
			"	Mn_5Si_3	D_{6h}^{3}	0.858	—	0.645	0.751	—	—	2	[201, 202]
	Gd_5Ge_4	26.97	Rhombic	Sm_5Ge_4	$Pn2_1a$	0.769	1.475	0.776	—	—	—	—	[441]
	GdGe	31.58	"	CrB	D_{2h}^{17}	0.4175	1.061	0.396	—	—	—	4	[199, 200]
			"	CrB	D_{2h}^{17}	0.432	1.079	0.398	—	—	—	4	[201, 203]
			"	CrB	D_{2h}^{17}	0.4327	1.077	0.3954	—	—	—	4	[434]
			"	CrB	D_{2h}^{17}	0.4339	1.0788	0.3973	—	—	—	4	[435]
	Gd_2Ge_3	40.91	Tetragonal	—	—	1.093	1.667	—	—	—	—	—	[435]
			Hexagonal	AlB_2	D_{6h}^{1}	0.3973	—	0.4179	1.052	—	—	1	[201]
	$GdGe_2$	48.00	Tetragonal	α-$ThSi_2$	D_{4h}^{19}	0.412	—	1.372	3.330	—	—	4	[146, 201]

TABLE 1. (Continued)

Metal	Germanide	Germanium content, % by mass	Structure	Structural type	Space group	Lattice constants, nm — a	b	c	c/a	Density, g/cm³ — X-ray	Pycnometric	Number of formal units	References
Gd	$GdGe_{2-n}$	(at) 600°C	Rhombic	$\alpha GdSi_2$	D_{2h}^{28}	0.4130	0.4096	1.376	—	—	—	4	[201]
Tb	Tb_5Ge_3	21.51	Hexagonal	Mn_5Si_3	D_{6h}^{3}	0.849	—	0.632	0.751	—	—	2	[202]
	Tb_5Ge_4	26.76	Rhombic	Sm_5Ge_4	$Pn21a$	0.762	1.466	0.772	—	—	—	—	[441]
	$TbGe$	31.35	»	CrB	D_{2h}^{17}	0.429	—	1.069	3.94	—	—	4	[203]
			Tetragonal	—	—	1.089	—	1.649	—	—	—	—	[435]
			Rhombic	CrB	D_{2h}^{17}	0.4296	1.0709	0.3943	—	—	—	4	[434]
				CrB	D_{2h}^{17}	0.4300	1.0717	0.3950	—	—	—	4	[435]
	Tb_2Ge_3	40.66	Hexagonal	AlB_2	D_{6h}^{1}	0.395	—	0.416	1.053	—	—	1	[145, 146]
Dy	Dy_5Ge_3	21.14	Hexagonal	Mn_5Si_3	D_{6h}^{3}	0.8438	—	0.6336	0.750	—	—	2	[142]
			»	Mn_5Si_3	D_{6h}^{3}	0.842	—	0.632	0.751	—	—	2	[202]
	$DyGe$	30.88	Rhombic	CrB	D_{2h}^{17}	0.4112	1.081	0.3924	—	—	—	4	[199, 200]
				CrB	D_{2h}^{17}	0.426	1.065	0.393	—	—	—	4	[200, 203]
			Tetragonal	—	—	1.081	—	1.629	—	—	—	—	[435]
			Rhombic	CrB	D_{2h}^{17}	0.4263	1.0675	0.3924	—	—	—	4	[434]
				CrB	D_{2h}^{17}	0.4272	1.0678	0.3931	—	—	—	4	[435]
	Dy_2Ge_3	40.12	Hexagonal	AlB_2	D_{6h}^{1}	0.392	—	0.413	1.050	—	—	1	[145, 146]
Ho	Ho_5Ge_3	20.89	»	Mn_5Si_3	D_{6h}^{3}	0.8410	—	0.630	0.750	—	—	2	[202, 205]
	$HoGe$	30.56	Rhombic	CrB	D_{2h}^{17}	0.425	1.062	0.391	—	—	—	4	[203]
			Tetragonal	—	—	1.079	—	1.623	—	—	—	—	[435]
			Rhombic	CrB	D_{2h}^{17}	0.4234	1.0610	0.3911	—	—	—	4	[434]
				CrB	D_{2h}^{17}	0.4247	1.063	0.3919	—	—	—	4	[435]
	Ho_2Ge_3	39.76	Hexagonal	AlB_2	D_{6h}^{1}	0.390	—	0.411	1.053	—	—	1	[145, 146]
Er	Er_5Ge_3	20.66	»	Mn_5Si_3	D_{6h}^{3}	0.835	—	0.626	0.750	—	—	2	[202]
			»	Mn_5Si_3	D_{6h}^{3}	0.8367	—	0.6266	0.748	—	—	2	[205]
	Er_5Ge_4	25.77	Rhombic	Sm_5Ge_4	$Pn21a$	0.751	1.441	0.759	—	—	—	—	[441]
	Er_2Ge_3	39.43	Hexagonal	AlB_2	D_{6h}^{1}	0.389	—	0.409	1.051	—	—	1	[145, 146]

TABLE 1. (Continued)

Metal	Germanide	Germanium content, % by mass	Structure	Structural type	Space group	a	b	c	c/a	Density, g/cm³ X-ray	Density, g/cm³ Pyc-nometic	Number of formal units	References
Er	ErGe	30.27	Hexagonal	CrB	D_{2h}^{17}	0.4215	1.0567	1.3901	—	—	—	4	[434]
						0.420	1.058	0.392	—	—	—	4	[435]
Tu	Tu₅Ge₃	20.50	Tetragonal	—	—	1.076	—	1.609	—	—	—	—	[435]
	Tu₂Ge₃	39.19	Hexagonal	Mn₅Si₃	D_{6h}^{3}	0.831	—	0.623	0.750	—	—	2	[202]
	TuGe	38.62	Hexagonal	AlB₂	D_{6h}^{1}	0.388	—	0.407	1.048	—	—	1	[145, 146]
Yb	YbGe₃	41.15	Tetragonal	—	—	1.072	—	1.601	—	—	—	—	[435]
	Yb₃Ge₅		Hexagonal	AlB₂	D_{6h}^{1}	0.3960	—	0.418	1.055	—	—	1	[145, 146]
	YbGe		Hexagonal	Th₃Pt₅	$P\bar{6}2m$	0.6803	—	0.4166	0.612	8.77	—	—	[206]
Lu	Lu₅Ge₃	19.93	Hexagonal	Mn₅Si₃	D_{6h}^{3}	0.824	—	0.617	0.749	—	—	2	[435]
	Lu₂Ge₃	38.36	Hexagonal	AlB₂	D_{6h}^{1}	0.383	—	0.405	1.057	—	—	1	[202]
	LuGe		Tetragonal	—	—	1.063	—	1.578	—	—	—	—	[145, 146]
Ti	Ti₃Ge	33.56	Hexagonal	Fe₃P	S_{4}^{2}	1.029	—	0.514	0.50	—	—	8	[435]
	Ti₅Ge₃	47.63	Hexagonal	Mn₅Si₃	D_{6h}^{3}	0.7537	—	0.5223	0.693	—	—	2	[112]
	TiGe	60.25	Rhombic	TiSi	C_{2v}^{1}	0.3807	0.5230	0.6833	—	5.86	—	4	[105, 106]
	TiGe₂	75.19	Rhombic	TiSi₂	D_{2h}^{24}	0.8577	0.5020	0.8846	—	6.69	—	4	[211]
Zr	Zr₃Ge	20.97	Hexagonal	—	—	0.8140	—	0.7170	0.880	—	—	5	[82]
	Zr₅Ge₃	32.20	Tetragonal	Ti₃P	C_{4h}^{4}	1.108	—	0.548	0.490	—	—	8	[214]
			Hexagonal	Mn₅Si₃	D_{6h}^{3}	0.7993	—	0.5597	0.700	—	—	2	[112]
	Zr₅Ge₃		Hexagonal	Mn₅Si₃	D_{6h}^{3}	0.799	—	0.554	0.693	7.29	7.12	2	[87]
	Zr₅Ge₄	38.90	Rhombic	Zr₅Si₄	—	0.724	—	1.316	1.810	—	—	2	[170, 214]
	ZrGe	44.20	—	FeB	D_{2h}^{16}	0.7075	0.3904	0.5396	—	—	—	—	[112]
	ZrGe₂		Rhombic	ZrSi₂	D_{2h}^{17}	0.3804	1.501	0.3764	—	—	—	4	[426]
		61.41	Rhombic	ZrSi₂	D_{2h}^{17}	0.3789	1.4975	0.3706	—	—	—	4	[82]
												4	[215]
Hf	Hf₃Ge	11.94	Tetragonal	Ti₃P	C_{4h}^{4}	1.092	—	0.542	0.496	—	—	8	[112]

TABLE 1. (Continued)

Metal	Germanide	Germanium content, % by mass	Crystal-chemical properties — Structure	Structural type	Space group	Lattice constants, nm — a	b	c	c/a	Density, g/cm³ — X-ray	Pyc-nometic	Number of formal units	References
Hf	Hf_2Ge	16.89	Tetragonal	$CuAl_2$	D_{4h}^{18}	0.6574	—	0.5361	0.815	—	—	4	[86]
	Hf_5Ge_3	19.60	Hexagonal	Mn_5Si_3	D_{6h}^{3}	0.7855—0.7895	—	0.5546—0.5534	0.706—0.701	—	—	2	[86]
	Hf_3Ge_2	21.33	»	Mn_5Si_3	D_{6h}^{3}	0.7883	—	0.5537	0.702	—	—	2	[108]
	$HfGe$	28.90	Tetragonal	U_3Si_2	D_{4h}^{5}	0.706	—	0.372	0.526	—	—	2	[112]
	$HfGe$		Rhombic	FeB	D_{2h}^{16}	—	1.5004	—	—	—	—	—	[86]
	$HfGe_2$	44.84	»	$ZrSi_2$	D_{2h}^{17}	0.3815	—	0.3779	—	—	—	4	[215]
V	V_3Ge	32.20	Cubic	β-W	O_h^{3}	0.4759	—	—	—	—	—	2	[82, 112]
	V_5Ge_3	46.09	Hexagonal	Mn_5Si_3	D_{6h}^{3}	0.7262—0.7250	—	0.4953—0.4967	0.682—0.685	—	—	2	[111]
	V_5Ge_3		Tetragonal	W_5Si_3	D_{4h}^{18}	0.955	—	0.483	0.506	—	—	4	[111]
	V_3Ge_2	48.71	Rhombic	Cr_3Ge_2	—	1.341	1.609	0.502	—	—	—	—	[111]
	$V_{11}Ge_8$	50.89	Rhombic	Mn_5Si_3	—	—	—	—	—	6.99	6.89	—	[216]
	VGe_2	74.02	»	—	—	—	—	—	—	—	—	—	[111]
	$V_{17}Ge_{31}$	80.06	Complex crystalline	$V_{17}Ge_{31}$	—	0.590 (0.590)	—	8.367 (0.492)	14.181 (0.8339)	—	—	—	[216, 217, 481]
Nb	Nb_3Ge	20.66	Cubic	β-W	O_h^{3}	0.5168	—	—	—	8.47	8.17	2	[113]
	Nb_5Ge	28.09	»	β-W	O_h^{3}	0.5167—0.5177	—	—	—	8.64	8.70	2	[114]
	α-Nb_5Ge_3	31.88	Tetragonal	Mo_5Si_3	D_{4h}^{18}	1.0148	—	0.5152	0.508	—	—	4	[82, 109]
	β-Nb_5Ge_3	34.25	Hexagonal	Mn_5Si_3	D_{6h}^{3}	0.7718	—	0.5370	0.696	—	—	2	[86, 219]
	$NbGe_2$	60.98	»	$CrSi_2$	D_6^{4}	0.4957	—	0.6770	1.366	—	—	3	[219]
			»	$CrSi_2$	D_6^{4}	0.4966	—	0.6781	1.365	8.20	7.81	3	[82]
			»	$CrSi_2$	D_6^{4}	0.4943	—	0.6778	1.371	8.295	8.17	3	[113]
Ta	Ta_3Ge (в)	11.79	Tetragonal	Fe_3P	S_4^{2}	1.036	—	0.516	0.50	—	—	8	[72]
	Ta_3Ge (н)		»	Ti_3P	C_{4h}^{4}	1.028	—	0.522	0.51	—	—	8	[112]
	Ta_2Ge	16.71	—	—	—	—	—	—	—	—	—	—	[112]
	α-Ta_5Ge_3	19.40	Tetragonal	Ta_5Si_3	D_{4h}^{18}	0.6599	—	1.201	1.820	—	—	4	[109, 110], [219]

TABLE 1. (Continued)

Metal	Germanide	Germanium content, % by mass	Structure	Structural type	Space group	a	b	c	c/a	Density X-ray	Density Pycnometric	Number of formal units	References
Ta	β-Ta₅Ge₃		Tetragonal	Mo₅Si₃	D_{4h}^{13}	1.001	—	0.5150	0.514	—	—	4	[219]
	Ta₅Ge₃ (5%C)		Hexagonal	Mn₅Si₃	D_{6h}^{1}	0.7581	—	0.5235	0.690	—	—	2	[87]
	TaGe₂	44.51	»	CrSi₂	D_{6}^{4}	0.4948	—	0.6737	1.362	11.392	11.28	3	[82]
Cr	Cr₃Ge	31.76	Cubic	β-W	O_{h}^{3}	0.4614	—	0.6738	1.363	7.68	7.28	2	[72]
	Cr₅Ge₃	45.58	Tetragonal	Cr₅Si₃	D_{4h}^{18}	0.9413	—	0.4780	0.507	—	—	4	[82]
	Cr₃Ge₂	48.31	Rhombic	Cr₃Si₂	—	—	—	—	—	—	—	—	[87]
	Cr₁₁Ge₈	50.38	Rhombic	Mn₅Si₃	D_{2h}^{16}	1.315	1.575	0.494	—	7.48	7.47	4	[82]
	CrGe	58.26	Cubic	FeSi	T^{4}	0.4780	—	—	—	—	—	4	[216]
	Cr₁₁Ge₁₉ (CrGe₂)	70.69	Tetragonal	Mn₁₁Si₁₀	D_{2d}^{8}	0.580 (0.580)	—	5.234 (0.476)	9.02 (0.8207)	—	—	—	[82]
	CrGe₃	80.72	—	—	—	—	—	—	—	—	—	—	[216, 481]
Mo	Mo₃Ge	20.14	Cubic	β-W	O_{h}^{3}	0.4933	—	—	—	9.97	9.70	2	[117]
	Mo₅Ge₃	31.22	Tetragonal	W₅Si₃	D_{4h}^{18}	0.9837	—	0.4973	0.606	—	—	4	[221]
	Mo₃Ge₂	33.53	»	Mo₃Si	—	—	—	—	—	—	—	—	[224]
	Mo₂Ge₃	53.16	—	—	—	—	—	—	—	—	—	—	[118]
	Mo₁₃Ge₂₃	57.24	»	Mo₁₃Ge₂₃	D_{2d}^{8}	0.599 (0.599)	—	6.354 (0.489)	10.607 (0.8164)	8.52	8.56	4	[118]
	MoGe₂	60.20	Rhombic	PbCl₂	D_{2h}^{16}	0.6343	0.3451	0.8582	—	—	—	—	[216, 481]
	α-MoGe₂		—	—	—	—	—	—	—	—	—	—	[225]
	β-MoGe₂	56.26	Tetragonal	MoSi₂	D_{4h}^{17}	0.3313	—	0.8195	2.474	8.91	—	2	[118]
	MoGe₁,₇	56.26	Hexagonal	Mg₃Cd	$I4_{1}22$	0.5994	—	4.3995	7.339	—	6.90	2	[118]
Mn	Mn₃,₂₅Ge	28.91	Hexagonal	NiAs	D_{6h}^{4}	0.5336	—	0.4365	0.816	—	—	2	[119]
	Mn₅Ge₂	34.58	»	—	D_{6h}^{4}	—	—	—	—	—	—	—	[119]
	Mn₅Ge₃	44.22	»	Mn₅Si₃	D_{6h}^{3}	0.7184	—	0.5053	0.703	—	—	2	[136, 137]
	Mn₃Ge₂	46.84	—	—	—	—	—	—	—	—	—	—	[119]
	Mn₁₁Ge₈	49.01	Rhombic	Mn₁₁Si₃	—	1.322	1.583	0.509	—	7.39	7.38	—	[216]

TABLE 1. (Continued)

| Metal | Germanide | Germanium content, % by mass | Structure | Structural type | Space group | Lattice constants, nm | | | | Density, g/cm³ | | Number of formal units | References |
						a	b	c	c/a	X-ray	Pyc-nome-tric		
Re	$ReGe_2$	43.79	Complex crystalline	—	—	—	—	—	—	—	—	—	[222]
Fe	$Fe_2Ge(Fe_{1.76}Ge)$	39.39	Hexagonal	NiAs	D_{6h}^4	0.4038	—	0.5027	1.246	—	—	2	[135–137]
	$(Fe_{1.7}Ge)$		»	Ni_2In	D_{6h}^4	0.4017	—	0.5004	1.246	—	—	2	[122, 124]
	Fe_5Ge_3	43.82	»	NiAs	D_{6h}^4	0.3978–0.4034	—	0.4993–0.5024	1.255–1.245	—	—	2	[122, 124]
	$Fe_{3.25}Ge$	28.57	»	Mg_3Cd	D_{6h}^4	0.5182–0.5188	—	0.4227–0.4234	0.815–0.816	8.10	—	2	[121]
	Fe_3Ge	30.23	Cubic	Cu_3Au	O_h^1	0.3574	—	—	—	—	—	1	[122, 124]
	Fe_4Ge_3	49.36	Complex crystalline	—	—	—	—	—	—	—	—	—	[122, 124]
	$FeGe$	56.52	Hexagonal	—	D_{6h}^1	0.4997	—	0.4055	0.811	—	—	—	[122, 124]
			»	—	D_{6h}^1	0.5003	—	0.4055	0.815	—	—	—	[127]
	$FeGe_2$	72.22	Tetragonal	$CuAl_2$	D_{4h}^{18}	0.5899	—	0.4941	0.838	7.70	—	4	[126, 238]
			»	$CuAl_2$	D_{4h}^{18}	0.5906	—	0.4950	0.838	—	—	4	[126]
			»	$CuAl_2$	D_{4h}^{18}	0.5909	—	0.4955	0.838	—	—	4	[244]
Co	Co_2Ge	38.12	Hexagonal	NiAs	D_{6h}^4	0.3917	—	0.5045	1.285	—	—	2	[130]
	$(Co_{1.76}Ge)$	41.16	»	NiAs	D_{6h}^4	0.3932	—	0.5014	1.275	—	—	2	[170]
	$(Co_{1.7}Ge)$	42.01	»	NiAs	D_{6h}^4	0.3917	—	0.4979	1.271	—	—	2	[136]
	$CoGe$	55.19	Monoclinic	Ni_3Sn_4	C_{2h}^3	1.1648	0.3807	0.4945	$[\beta=1.77\ rad(\beta=101.1°)]$	8.24	—	16.2	[131]
	Co_5Ge_3	64.88	—	—	—	—	—	—	—	—	—	—	[130]
	Co_5Ge_7	63.29	—	—	C_{4v}^9	0.7641	—	0.5814	0.760	—	—	—	[132]
	Co_5Ge_8	66.33	Tetragonal	—	—	0.7641	—	0.5814	0.7608	—	—	—	[131]
	$CoGe_2$	71.12	Face-centered rhombic, pseudotetragonal	$CoGe_2$	D_{2o}^{17}	0.5681	0.5681	1.0811	1.90	—	—	8	[130, 170]
Ni	Ni_3Ge	29.19	Cubic	Cu_3Au	O_h^1	0.3567	—	—	—	—	—	1	[134]
	$Ni_2Ge(Ni_{1.7}Ge)$	38.21	Hexagonal	NiAs Ni_2In	D_{6h}^4	0.3919	—	0.5046	1.287	—	—	2	[136, 137]
			»	NiAs, Ni_2In	D_{6h}^4	0.3848	—	0.4921	1.279	—	—	2	[129]
	$Ni_{1.86}Ge$		»	NiAs, Ni_2In	D_{6h}^4	0.3954	—	0.5045	1.275	—	—	2	[129]

TABLE 1. (Continued)

Metal	Germanide	Germanium content, % by mass	Crystal-chemical properties										
			Structure	Structural type	Space group	Lattice constants, nm				Density, g/cm³		Number of formal units	References
						a	b	c	c/a	X-ray	Pycnometric		
Ni	NiGe	55.29	Rhombic	MnP	D_{2h}^{16}	0.5810	0.5380	0.3427	—	—	—	4	[134]
Ru	RuGe	41.65	Cubic	FeSi	T^4	0.4846	—	—	—	—	—	4	[138]
	Ru₂Ge₃	51.71	Tetragonal	Ru₂Si₃	—	0.5709	—	0.4650	0.815	—	—	—	[138]
	RuGe₂	58.94	»	RuSi₂	—	—	—	—	—	—	—	—	[82]
Rh	Rh₂Ge	26.06	Rhombic	PbCl₂	D_{2h}^{16}	0.544	0.757	0.400	—	11.20	11.40	4	[83]
	Rh₅Ge₃	29.73	»	—	D_{2h}^9	0.542	1.032	0.396	—	11.0	11.6	2	[83]
	RhGe	41.36	»	MnP	D_{2h}^{16}	0.570	0.648	0.325	—	9.70	9.70	4	[83]
	Rh₃Ge₄	48.46	Tetragonal	—	D_{2d}^{12}	0.570	—	1.00	1.754	—	8.50	—	[252]
	Rh₁₇Ge₂₂	47.71	»	Rh₁₇Ge₂₂	—	0.560	—	7.845	13.998	—	—	—	[479]
Pd	Pd₄Ge	14.53	—	—	—	—	—	—	—	—	—	—	[134, 254]
	Pd₅Ge₂	21.39	—	—	—	—	—	—	—	—	—	—	[139]
	Pd₂Ge	25.39	Hexagonal	Fe₂P	D_3^2	0.6673	—	0.3386	0.507	—	—	3	[139, 254]
	PdGe	40.49	Rhombic	MnP	D_{2h}^{16}	0.6258	0.5781	0.3481	—	—	—	4	[134, 139]
	Pd₈₄Ge₁₆	11.47	Cubic	β-W	O_h^3	0.3137	—	—	—	—	—	2	[255]
Os	OsGe₂	43.28	Tetragonal	RuSi₂	C_{2h}^3	0.8995	0.3094	0.7685	[β=2.08 rad (β=119.1°)]	11.90	11.10	4	[140]
Ir	IrGe	27.41	Rhombic	MnP	D_{2h}^{16}	0.6280	0.5611	0.3490	—	—	—	4	[131, 134]
	Ir₄Ge₅	32.07	Tetragonal	Ir₄Ge₅	D_{2d}^6	0.564	—	1.824	3.234	—	—	4	[131, 479]
	Ir₃Ge₇	46.85	Cubic	Ir₃Sn₇	O_h^9	0.8752	—	—	—	—	—	4	[131, 139]
	IrGe₄	65.38	Hexagonal	—	—	0.6211	—	0.7772	1.251	—	—	4	[131]
Pt	Pt₃Ge	11.03	Monoclinic	—	C_{2h}^3	0.7930	0.7767	0.5520	[β=0.76 rad (β=44.72°)]	—	—	1	[131]
			Quasicubic	Cu₃Au	O_h^1	0.7931	0.7767	0.7767	[β=1.57 rad (β=90.06°)]	—	—	1	[131]
	Pt₂Ge	15.68	Hexagonal	Fe₂P	D_3^2	0.6683	—	0.3527	0.528	—	—	3	[131, 139]
	Pt₃Ge₂	19.87	Rhombic A, face-centered (superstructure)	MnP	D_{2h}^{16}	0.7544	0.3423	1.2236	—	—	—	4	[131]
	PtGe	27.11	Rhombic	MnP	D_{2h}^{16}	0.6088	0.5732	0.3711	—	—	—	4	[134]
	Pt₂Ge₃	35.81	»	MnP	D_{2h}^{16}	1.6430	0.3378	0.6221	—	—	—	4	[131, 134]

TABLE 1. (Continued)

Metal	Germanide	Germanium content, % by mass	Crystal-chemical properties — Structure	Structural type	Space group	Lattice constants, nm — a	b	c	c/a	Density, g/cm³ — X-ray	Pycnometric	Number of formal units	References
Pt	$PtGe_2$	42.66	Rhombic	$CaCl_2$	D_{2h}^{12}	0.6185	0.5767	0.2908	—	—	—	2	[131]
Th	Th_3Ge	9.44	—	—	—	—	—	—	—	—	11.0	—	[150]
	Th_3Ge_2	17.26	Tetragonal	Th_3Si_2	D_{4h}^{5}	0.7971	—	0.4170	0.523	—	10.48	2	[148]
	Th_3Ge_2		»	Th_3Si_2	D_{4h}^{5}	0.7951	—	0.4194	0.527	10.54	—	2	[150]
	$ThGe$	23.83	Cubic	$NaCl$	O_h^{5}	0.6033	—	—	—	9.17	9.15	4	[148]
	$ThGe$		»	$NaCl$	O_h^{5}	0.6046	—	—	—	9.45	9.44	4	[150]
	$ThGe_{1.5}$	31.93	Distorted rhombic	AlB_2	D_{6h}^{1}	0.6989	0.8432	0.8136	—	9.34	—	—	[150]
	$ThGe_{1.6}$	33.35	The same	$α\text{-}ThSi_2$	D_{4h}^{19}	0.5913	0.5889	1.4219	—	—	—	4	[150]
	$α\text{-}ThGe_2$	38.48	Tetragonal	$β\text{-}ThSi_2$	D_{4h}^{19}	0.4106	—	1.4193	3.456	—	—	4	[148]
	$β\text{-}ThGe_2$		Distorted hexagonal	AlB_2	D_{6h}^{1}	—	—	—	—	—	—	4	[148]
	$ThGe_2$	38.48	Rhombic	$ZrSi_2$	D_{2h}^{17}	0.4223	1.6911	0.4052	—	8.66	8.64	4	[149, 150]
	$Th_{0.9}Ge_2$	41.00	»	—	D_{2h}^{19}	1.6642	0.4023	0.4160	—	8.44	—	—	[149, 150]
	$ThGe_3$	48.42	Pseudocubic	—	—	1.172	—	—	—	—	—	4	[148]
U	U_3Ge	4.17	—	—	—	—	—	—	—	—	13.40	2	[84]
	U_5Ge_3	15.47	Hexagonal	Mn_5Si_3	D_{6h}^{3}	0.8577	—	0.5791	0.675	—	—	—	[57]
	U_3Ge_4	28.90	Rhombic	—	—	0.5871	0.9879	0.8977	—	—	—	4	[257]
	UGe_2	37.88	»	$ZrSi_2$	D_{2h}^{17}	0.4118	1.5130	0.3978	—	—	—	4	[257]
	UGe_3	47.78	Cubic	Cu_3Au	O_h^{1}	0.4198	—	—	—	—	—	1	[197]
			»	Cu_3Au	O_h^{1}	0.4196	—	—	—	10.245	—	1	[360]
			»	Cu_3Au	O_h^{1}	0.4206	—	—	—	10.370	—	1	[256. 257]
Pu	Pu_4Ge_3	30.85	Hexagonal	AlB_2	D_{6h}^{1}	0.3975	—	0.4198	1.056	10.06	—	1/2	[259]
	$PuGe_2$	37.30	Tetragonal	$α\text{-}ThSi_2$	D_{4h}^{19}	0.4102	—	1.3810	3.360	10.98	—	4	[259]
	$PuGe_3$	47.16	Cubic	Cu_3Au	O_h^{1}	0.4223	—	—	—	10.07	—	1	[259]
N	Ge_3N_4	79.54	Rhombic	Ge_3N_4	C_{3i}^{2}	1.384	0.906	0.8180	—	—	—	6	[305]
	Ge_3N_4		Rhombohedral	Ge_3N_4	C_{3i}^{2}	0.8587	—	—	$[α=1.88$ rad $(α=107°)]$	5.29	5.25	6	[303, 304]

TABLE 1. (Continued)

Metal	Germanide	Germanium content, % by mass	Structure	Structural type	Space group	Lattice constants, nm a	b	c	c/a	Density, g/cm³ X-ray	Pycnometric	Number of formal units	References
As	$GeAs_2$	32.65	Rhombohedral	Ge_3N_4	C_{3i}^2	0.8620	—	—	[α=1.89 rad (α=108°)]	—	—	6	[222, 427]
			Rhombic	—	D_{2h}^9	0.3721	1.0120	1.4739	—	—	—	—	[308]
	GeAs	49.21	»	—	D_{2h}^9	1.4760	1.0160	0.3728	—	—	—	—	[307]
			Monoclinic	—	—	2.2124	0.3777	0.9448	[β=0.77 rad (β=43.97°)]	—	—	—	[308]
O	GeO_2 Insoluble	69.41	Tetragonal	Rutile	—	0.4403	—	0.2865	0.6513	6.26	6.239	—	[19, 170]
	GeO_2 Soluble		Hexagonal	Cristobalite	—	0.4972	—	0.5648	1.135	4.28	4.228	—	[19, 170]
S	GeS_2	53.10	Rhombic	GeS_2	C_{2v}^{19}	1.1683	2.2385	0.6783	—	—	3.01	24	[411]
	GeS	69.36	»	GeS	D_{2h}^{16}	0.4298	1.0440	0.3647	—	—	4.012	4	[411]
Se	$GeSe_2$	31.50	»	—	—	1.2985	0.6943	2.2134	—	—	4.56	—	[332]
			»	—	—	0.6952	1.2220	2.3035	—	—	—	—	[333]
	GeSe	47.90	Tetragonal	—	—	0.8847	—	0.9779	1.105	—	5.30	—	[332]
			Rhombic	—	D_{2h}^{16}	0.4383	0.3832	1.0821	—	—	—	—	[334]
Te	GeTe	36.26	Distorted tetragonal	NaCl	—	0.5980	—	—	[α=1.546 rad (α=88.35°)]	6.29	6.20	4	[336]
	GeTe (at 460°C)	36.26	Cubic	NaCl	O_h^5	0.601	—	—	—	—	—	4	[339]
	GeTe (at 390°C)		»	NaCl	O_h^5	0.5992	—	—	—	—	—	4	[339]
	Ge_2Te_3		Hexagonal	—	D_{3d}^3	0.432	—	5.30	12.27	—	—	—	[339]

h) High-temperature modification; 1) low-temperature modification.

TABLE 2. Physical Properties of Transition-Metal Germanides

Compound	Heat of formation kJ/mole	Heat of formation kcal/mole	Enthalpy of formation kJ/mole	Enthalpy of formation kcal/mole	Enthalpy decomposition kJ/mole	Enthalpy decomposition kcal/mole	Melting or decomposition point, °C	Resistivity 25 °C	Resistivity -196 °C	Temperature coefficient of electrical resistance, deg⁻¹·10³	Coefficient of thermal conductivity, W/cm·deg	Temperature of transition to superconductive state, °K	Coefficient of thermal emf, µV/deg	Hall's constant R·10⁴, cm³/C	Curie point, °C	Modulus of elasticity MN/m²	Modulus of elasticity kg/mm²	Microhardness MN/m²	Microhardness kg/mm²
Ti₅Ge₃							>2000•	163.9•				<1.20 [109, 110]	-6.5•	-30.2•				13000	1300 [464]
TiGe							>1400•	152.2•				<1.20 [109, 110]	-4.9•	2.0•				12180	1218•
TiGe₂							~1200•	1280 [443]					-4.2	60 [443]				4350	435 [464]
Zr₅Ge₃							1585 [214]	142.0•				0.6•	1.0•					10130	1015•
Ar₅Ge₃							2330 [214]	160.0•				<0.03 [405]						11260	1126 [464]
ZrGe							2240 [214]	61.8•					-8.2•	-39.3•				9100	910•
ZrGe₂							1520 [214]	126 / 410 [68]		3.5 [68]		<0.03 [110, 405]	-5.6•	-23.8 [68]				6800 / 8800	680 [68] / 880 [464]
V₃Ge							>1700•	89 / 218				6.01 [110]	10 [218]	64•				15400	1540 [464]
V₅Ge₃							>1800•	157.1•					-3.3•	0.3				14150	1415 [464]
V₃Ge₂							~1600•	195.0					2.9•	-1.0•					
VGe₂							~1200•	1160 [443]						5.3 [443]				11400•	1140•
Nb₃Ge	-111.3	-26.6 [114]			448	116.6 [114]	>1900 [113]					6.3-6.9 [12, 405]							
α-Nb₅Ge₃	-113.4	-27.1 [118]			512.04	122.3 [115]	>1900 [219]	118.2•				<1.02 [405]	-3.6•	-4.4•				15880	1588•
β-Nb₅Ge₃	-95.8	-22.9 [115]					>1600 [219]												
NbGe₂	-43.5	-10.4 [115]						67 [72] / 118 [68]	31 [72]	2.9 [68]	0.31 [72]	<1.2 [110, 285]	12 [72] / 12	7 [68]				7450	745 [68]
Ta₂Ge							~1300•	330.0•				1.60 [110]	-5.4•	-12.2•				16400	1640•
α-Ta₅Ge₃							>1700•	159.9•				<1.20 [110]	-9.3•	-6.1•				16000	1600 [464]
β-Ta₅Ge₃							>1700•												
TaGe₂							~1300•	35 [72] / 70.5•	13 [72]		0.35 [72]	<1.20 [110, 405]	11 [72]	0.7•				13800	1380•
Cr₃Ge							1520 [85]	30.5 [116]					10.9 / 13.0 [220]	0.37•				9400 / 13700	940 [116] / 1370 [464]

TABLE 2. (Continued)

Compound	Heat of formation kJ/mole	Heat of formation kcal/mole	Enthalpy of formation kJ/mole	Enthalpy of formation kcal/mole	Enthalpy decomposition kJ/mole	Enthalpy decomposition kcal/mole	Melting or decomposition point, °C	Resistivity, µohm·cm 25 °C	Resistivity, µohm·cm -196 °C	Temperature coefficient of electrical resistance, deg^{-1}·10^3	Coefficient of thermal conductivity, W/cm·deg	Temperature of transition to superconductive state, °K‡	Coefficient of thermal emf, µV/deg	Hall's constant R·10^4, cm^3/C	Curie point, °C	Modulus of elasticity MN/m²	Modulus of elasticity kg/mm²	Microhardness MN/m²	Microhardness kg/mm²
CrGe							>1750 [118]	201.0•				<1.20 [109, 110]	-0.4•	0.08•				13000-14700	1300-1470 [220]
Mo$_3$Ge	60.7	14.5 [226]					1780-1800 [85]	33.5-40 [116]	[116]			1.43 [109, 110]	-5.1•	0.1•				11450	1145 [116]
Mo$_3$Ge$_2$																			
MoGe$_2$							~900 [119]	70.5•				1.20 [109, 110]	0.5•	-2.0•				12500	1250 [464]
Mo$_{3.25}$Ge								250 [68] 153.0•		3.2 [68]		<1.20 [110]	1.4 [68]	-11.0 [68]		66	6.6 [227]	9400 9000	940 [68] 900 [464]
Mn$_3$Ge$_2$							920 [119]	~42 [235]											
Mn$_5$Ge$_3$							920 [119]								47 [136, 137]				
Mn$_3$Ge$_2$							745 [119]	~54 [235]							10 [236]				
ReGe$_2$	8.37	2 [222]					1132 [222]	~137 [235]					-4.0 [126]						
Fe$_2$Ge							1170 [126]								20 [126]				
Fe$_{3.25}$Ge							1220 [126]	20-30 [126]											
FeGe												<1.02 [405]	‡		530-550 [126]			5300	530 [126]
FeGe$_2$							866 [126]	10 [126]				<0.30 [405]	-0.4 [126]		67 127 [126]			8700	870 [127]
Co$_2$Ge							1200 [130]												
CoGe							982† [130]												
Co$_2$Ge$_3$							750† [130]												
CoGe$_2$							842† [130]												
CoGe$_3$																			
Ni$_3$Ge							1161† [135]												
Ni$_2$Ge							1200 [135]												
NiGe							850† [135]												

TABLE 2. (Continued)

Compound	Heat of formation kJ/mole	Heat of formation kcal/mole	Enthalpy of formation kJ/mole	Enthalpy of formation kcal/mole	Enthalpy decomposition kJ/mole	Enthalpy decomposition kcal/mole	Melting or decomposition point, °C	Resistivity μohm·cm 25°C	Resistivity μohm·cm -196°C	Coefficient of thermal conductivity, W/cm·deg	Temperature coefficient of electrical resistance, deg⁻¹·10³	Temperature of transition to superconductive state, °K‡	Coefficient of thermal emf, μV/deg	Hall's constant R·10⁴, cm³/C	Curie point, °C	Modulus of elasticity MN/m²	Modulus of elasticity kg/mm²	Microhardness MN/m²	Microhardness kg/mm²
RuGe							~1500 [138]												
Ru₂Ge₃							~1500 [138]												
Rh₂Ge							<1400 [252]											7000	700 [252]
Rh₅Ge₃												2.12 [83, 404]						4750	475 [252]
RhGe							1300 [252]					<1.0 [83]						5800	580 [252]
Rh₃Ge₄																		5000	500 [252]
PdGe												<0.30 [405]							
IrGe												4.70 [404]							
ScGe₂												1.30–1.31 [207]							
Y₅Ge₃							1800 ± 90	130 ± 6			1.44 ± 0.15		−8.1 ± 0.80	1.55 ± 0.05				9870	987 ± 59
YGe							1650 ± 75	79 ± 4			4.00 ± 0.40	3.80 [207]	0.7 ± 0.07	2.99 ± 0.10				8360	836 ± 84
YGe₂							1525 ± 75	98 ± 5			0.98 ± 0.10		0.8 ± 0.08	10.25 ± 0.40				5400	540 ± 26
La₅Ge₃							1285 ± 65	169 ± 8			0.83 ± 0.08		−1.7 ± 0.17	−0.93 ± 0.05				3450	345 ± 17
LaGe							1350 ± 70	67 ± 3			3.03 ± 0.30		−3.9 ± 0.39	1.99 ± 0.10				5200	520 ± 19
LaGe₂							1380 ± 70	87 ± 4 / 659 [76]			1.07 ± 0.10	1.49 [207]	−10.0 ± 1.0	5.87 ± 0.20				5250 / 3750	525 ± 22 / 375 [76]
Ce₅Ge₃							1290 ± 65	234 ± 11			2.82 ± 0.28		1.7 ± 0.17	−7.4 ± 0.30				4170	417 ± 23
CeGe							1415 ± 70	124 ± 6			0.55 ± 0.06		−6.1 ± 0.61	−2.55 ± 0.10				5360	536 ± 21
CeGe₂							1465 ± 75	165 ± 13			0.93 ± 0.09		0.7 ± 0.07	−11.40 ± 0.45				5720	572 ± 28
Pr₅Ge₃							1420 ± 70	265 ± 13			0.21 ± 0.02		0.15 ± 0.02	1.69 ± 0.05				5350	535 ± 21
PrGe							1470 ± 75	164 ± 8			2.11 ± 0.21		0.65 ± 0.07	2.78 ± 0.10				5650	565 ± 26
PrGe₂							1540 ± 80	193 ± 9			1.20 ± 0.12		0.25 ± 0.03	1.51 ± 0.05				6450	645 ± 41
Nd₅Ge₃							1520 ± 80	181 ± 9			0.68 ± 0.07		−7.0 ± 0.70	2.06 ± 0.10				6120	612 ± 45
NdGe							1550 ± 80	111 ± 5			1.93 ± 0.19		−12.9 ± 1.29	5.44 ± 0.20				6130	613 ± 31
NdGe₂							1710 ± 90	120 ± 6			3.14 ± 0.31		−3.9 ± 0.39	−1.54 ± 0.05				7600	760 ± 47
Sm₅Ge₃							1260 ± 65	300 ± 15			2.13 ± 0.21		−5.0 ± 0.50	−1.97 ± 0.10				4180	418 ± 14

TABLE 2. (Continued)

Compound	Heat of formation		Enthalpy of formation		Enthalpy decomposition		Melting or decomposition point, °C	Resistivity, μohm·cm		Temperature coefficient of electrical resistance, $deg^{-1} \cdot 10^3$	Coefficient of thermal conductivity, W/cm·deg	Temperature of transition to superconductive state, °K‡	Coefficient of thermal emf, μV/deg	Hall's constant $R \cdot 10^4$, cm^3/C	Curie point, °C	Modulus of elasticity		Microhardness	
	$\frac{kj}{mole}$	$\frac{kcal}{mole}$	$\frac{kj}{mole}$	$\frac{kcal}{mole}$	$\frac{kj}{mole}$	$\frac{kcal}{mole}$		25 °C	-196 °C							$\frac{MN}{m^2}$	$\frac{kg}{mm^2}$	$\frac{MN}{m^2}$	$\frac{kg}{mm^2}$
Sm_5Ge_3							1260 ± 65	300 ± 15		2.13 ± 0.21			-5.0 ± 0.50	-1.97 ± 0.10				4180	418 ± 14
SmGe							1310 ± 65	225 ± 11		2.96 ± 0.30			-4.0 ± 0.40	4.70 ± 0.20				4600	460 ± 18
$SmGe_2$							1320 ± 65	255 ± 13		1.17 ± 0.12			2.20 ± 0.22	6.40 ± 0.25				5380	538 ± 19
$EuGe_2$							1350	127 ± 6		2.12 ± 0.21									
Th_3Ge							1400 † [150]												
Th_3Ge_2							~1800 [150]												
ThGe							1800 [150]												
$ThGe_{1.5}$							1700 [150]												
$ThGe_2$							600 † [150]												
$Th_{0.9}Ge_2$							1600 [150]												
U_5Ge_3							1670 [84]												
U_3Ge_4							~1400 † [84]												
UGe_2							~1420 † [84]						< 0.30 [405]						
UGe_3							1450 [84]												

† Decomposition.

‡ A "less than" sign indicates that superconductivity has not been detected above this temperature.

§ All the values for rare-earth germanides from Y_5Ge_3 to $EuGe_2$ have been taken from the data of V. M. Rud'.

Nowotny phases have a variable composition, the amount of the stabilizing element increasing within each group of the periodic table as the atomic number of the metallic component rises and as the number of the group to which this component belongs increases. The capacity for stabilization of Nowotny phases increases from oxygen to nitrogen and then on to boron and silicon.

The crystal chemistry of the germanides and silicides of rare-earth metals has been considered by Parthe [489]. This author discusses possible structural types and the conditions for their formation, demonstrating that the structural element of compounds of the RX and RX$_2$ types (where X is Si or Ge and R is a rare-earth metal) is a trigonal prism, which is found in all rare-earth silicides and germanides.

Table 1 presents a summary of structural data on the germanides of certain metals.

PHYSICAL PROPERTIES AND ELECTRONIC STRUCTURE

Almost no research has been done on the physical properties of germanides. The available data, which are summarized in Table 2, show that germanides are compounds with rather high melting points, reaching 2000-2400° C for the germanides of certain transition metals.

However, these compounds are materially less high-melting than carbides or even silicides. According to the available data, the resistivity of germanides is somewhat higher than that of silicides, although that of most transition-metal germanides is at the level characteristic of metallic and metalloid phases (several tens of μohm·cm). The temperature coefficient of electrical resistance also has values normal for metallic and metalloid compounds; the same is true of the thermal emf and Hall constant. Some germanides have high points of transition to the superconductive state (T_k), particularly complex niobium compounds, which are formed on the basis of Nb$_3$Ge by partial replacement of the germanium by aluminum, tin, or gallium (T_k is 6.3°K for Nb$_3$Ge, 12.6°K for Nb$_3$Ge$_{0.5}$Al$_{0.5}$, 7.3°K for Nb$_3$Ge$_{0.5}$Ga$_{0.5}$, 17.6-18.0°K for Nb$_3$Sn$_{1-x}$Ge$_x$, and 12.6°K for Nb$_3$Sn$_{0.5}$Ge$_{0.5}$ [12]).

Certain compounds of germanium with vanadium also have a high T_k, e.g., 6.01°K for V$_3$Ge and 14.0°K for V$_3$Ge$_{0.1}$Si$_{0.9}$, but such compounds containing both vanadium and germanium as V$_3$Ge$_2$Ni$_5$ and V$_5$Ge$_3$C$_{0.05}$ do not enter the superconductive state at temperatures above 1.02°K [12].

The special features of the physical properties of germanides discussed above can be preliminarily interpreted on the basis of theories of the electronic structure of compounds formed by germanium with other elements.

In order to consider the electronic structure of germanides, it is expedient to subdivide all the elements of the periodic table that form compounds with germanium into three groups:
1) elements whose isolated atoms have outer (valence) s-electrons, with the deeper-lying electron shells either completely filled or completely free (s-elements);
2) elements whose isolated atoms have outer s-electrons, with incompletely filled deeper-lying d and f shells (ds- and fds-elements);
3) elements whose isolated atoms have sp-electrons (sp-elements).

The first of these groups comprises the germanides of alkali and alkali-earth metals. beryllium, magnesium, and the elements of the copper and zinc subgroups.

Compounds of s-Metals with Germanium. Isolated germanium atoms are characterized by an s^{2}p^2 valence-electron configuration, which can be converted to the sp^3

configuration characteristic of diamond by an $s \rightarrow p$ transition, although this configuration is considerably less stable in terms of energy than the sp^3 configuration of diamond or even that of silicon (see p. 26). In considering compounds of germanium with alkali metals, it must be assumed that they are formed as a result of transfer of s-electrons from the metals to the germanium atoms. On passing over to the germanium, these s-electrons stabilize the sp^3 configuration of the germanium atom. This corresponds to formation of ionic bonds between alkali metals and germanium. On the other hand, germanium atoms are capable of mutual stabilization of their sp^3 configurations when they combine into covalent groups. The capacity for such mutual stabilization should obviously increase and result in formation of germanium-atom polyanions as the ionization potential of the alkali metal decreases, i.e., when stabilization is provided both by mutual overlapping of the sp^3 configurations and by the s-electrons of the alkali metal. Conversely, if an alkali metal has a high ionization potential, the possibility of transfer of s-electrons to the sp^3 configuration of germanium should be statistically lower, which can lead to formation of unusual polycations of alkali-metal atoms. It is as a result of this phenomenon that lithium and germanium form two compounds, Li_6Ge_2 and Li_4Ge [59], the data of Gladyshevskii and Kripyakevich [60] showing that the latter must be assigned the formula $Li_{15}Ge_4$. This compound has a structure with dense packing of atoms of different sizes. It can be represented as a packing of polyhedra $(GeLi_3^{(1)}Li_9^{(2)})$, which emphasizes the formation of structural elements (polycations) of lithium atoms, similar to the process in lithium silicides [61]. The other alkali metals form MeGe phases with germanium [3, 49, 62]. The polygermanide $NaGe_4$ is also known for sodium [48]. Similar phases have been found for potassium, rubidium, and cesium [49]. In this case, NaGe has a monoclinic structure, while all other germanide phases (MeGe and $MeGe_4$) crystallize with cubic structures. The formation of cubic structures is probably due to the substantial stabilization of the germanium sp^3 configuration by the s-electrons of the most readily ionized alkali-metal atoms. In other words, the cubic structure characteristic of ionic compounds between alkali-metal cations and germanium atom polyanions is produced. In accordance with the fact that the sp^3 configuration of germanium has a lower energy stability than that of carbon (diamond) or silicon, formation of complex polyanions is more difficult in this case and the maximum germanium content corresponds to the $MeGe_4$ composition (the maximum carbon content for carbides corresponds to MeC_{60}, while the maximum silicon content for silicides corresponds to $MeSi_8$).

As a result of this factor, the temperature at which alkali-metal germanides can be synthesized from their elements is also high, ranging from 600 to 1000°C.

As for the corresponding silicides, we must assume the existence of rather broad energy gaps and semiconductive properties for the MeGe and $MeGe_4$ phases [62, 64]; preliminary measurement of the electrical resistance of KGe [64] has confirmed this hypothesis. Similar relationships are also to be expected for formation of beryllium, magnesium, and alkali-earth germanides. However, no compounds of beryllium with germanium have been observed, which is probably due to the fact that the s^2 configuration of the isolated beryllium atom can be converted to an sp configuration by an $s \rightarrow p$ transition. This results in a sharp decrease in the number of s-electrons that could be transferred to germanium to stabilize its sp^3 configuration. A material role is also played by the considerable difference in principal quantum numbers, i.e., in the energy levels of the valence electrons of germanium $(4s^2 4p^2)$ on the one hand and beryllium $(2s^2)$ on the other.

As we move on to magnesium, the probability of $s \rightarrow p$ transitions decreases as a result of the drop in the energy stability of the sp configuration as the principal quantum number of the sp electrons increases. This factor and the smaller difference in energy levels increases the probability that magnesium will transfer s-electrons to germanium. The resultant compound is Mg_2Ge, which is similar in composition to the germanide of lithium, the most difficultly ionized alkali metal.

Proceeding to calcium, a vacant 3d level appears and the overall pattern of electron transitions acquires the form $d \rightleftharpoons s^2 \rightleftharpoons sp$, i.e., the probability of $s \rightarrow p$ transitions producing sp configurations is sharply reduced and both d and s bonding functions appear. This permits formation of calcium germanides (as a result of transfer of s electrons to the germanium atom and attraction of the germanium electrons to the d level of the calcium atom) and a variety of germanide phases, which have the compositions Ca_2Te, $CaGe$, $CaGe_2$, Ca_7Ge, and $Ca_{33}Ge$. The formation of unusual calcium-atom polycations is caused by the tendency of these atoms to form stable d-electron configurations.

The literature contains no information on strontium germanides, but their existence can be presumed; the variety of phase compositions should decrease from calcium to strontium and from strontium to barium, since $s \rightarrow d$ transitions become more pronounced as the principal quantum number increases and there is no need of combination for formation of stable sp configurations. On this basis, it can be assumed that the transfer of germanium electrons to the d levels of the metals should increase as we move from calcium to barium, a phenomenon that should, in general, produce a decrease in forbidden-zone width in these compounds, which undoubtedly have semiconductive properties.

In the copper subgroup, germanium forms compounds only with copper itself, while silver and gold do not form germanides. Since isolated atoms of all the elements in this subgroup have a $d^{10}s^1$ electron configuration and are thus capable of transferring one s-electron to germanium and hence stabilizing its sp^3 configuration, the formation of copper germanides is understandable. The fact that there are no silver or gold germanides is due to the appearance in the metal atoms of completely vacant 4f and 5f levels, into which some of their d-electrons can be transferred (disrupting the d^{10} configuration), while the vacancies produced in the d shell can be statistically filled by the s-electrons, which greatly reduces the probability of their transfer to germanium atoms.

None of the metals of the zinc subgroup form germanide phases. Isolated atoms of the metals in this subgroup have a $d^{10}s^2$ valence-electron configuration. It is hard to imagine that the electrons could be transferred in s^2 pairs. Such transfer would require strong excitation of the s^2 electrons, which has not been achieved in the experiments that have been performed on the interaction of zinc with germanium. At the same time, the severe structural distortion of zinc during dissolution of germanium in it (see p. 56) shows that such an interaction, in severely limited form, begins in this case and it can be surmised that formation of zinc germanides is not impossible in principle. Moreover, the same process is probable for the germanium–mercury systems, thus casting doubt on their phase diagrams, which, according to the data in the literature [65-67], have a eutectic character. It is possible, however, that germanides of zinc-subgroup metals, which can theoretically be formed under very severe excitation, have an essentially metastable character and do not persist under normal conditions.

Consequently, the germanides of s-metals generally exhibit an ionic-covalent type of atoms chemical bonding, i.e., principally ionic between metal and germanium atoms or groups of atoms and principally covalent within groups of metal or germanium atoms. As a result, all or most of these compounds should have semiconductive properties, relatively high heats of formation (due to the energy stability of the covalently bonded groups of metal and germanium atoms), and the ability to undergo thermal dissociation (disproportionation) in the solid and particularly the gaseous states. The latter property should become stronger as the polycationic and polyanionic groups become more complex.

Compounds of ds- and fds-Metals with Germanium. In considering the electronic structure of transition-metal germanides, it is first necessary to note the analogy between germanides and silicides, which follows from the formal composition of the former, as well as from the similarity in the physical properties and crystal structures of transition-metal germanides and silicides [68]. This is obviously due to the fact that germanium and silicon have the same electronic structures (sp^3 valence-electron configurations) and to the closeness of the valence-electron energy levels of these elements. It is manifested in the rather small difference in forbidden-zone width [2, 8] [Si = $1.728 \cdot 10^{-19}$ J (1.08 eV) and Ge = $(1.04-1.232) \cdot 10^{-19}$ J (0.65-0.77 eV)], while the valence-electron evergy levels of germanium and silicon differ greatly from those of the substantially more stable sp^3 configuration of carbon (diamond) and from the substantially less stable sp^3 configuration of gray tin.

In considering the formation of transition-metal germanides, it is necessary to take into account the tendency of transition-metal atoms to accept or donate electrons to form stable configurations [69]. In this connection, it is best to discuss separately the possible electronic structures of transition metals whose isolated atoms have a d shell containing a number of electrons $N_d \leq 5$ and of transition metals with $N_d > 5$.

The metals of the first group are characterized by d^0 and d^5 stable configurations, the statistical weight of the d^5 configuration increasing with the number of electrons in the d shell of an isolated atom. According to the data yielded by x-ray spectral analysis, this corresponds to an increase in the number of localized electrons (those participating in covalent bonding) and a decrease in the relative number of shared electrons (those taking part in metallic bonding). Thus, according to the data of Korsunskii and Genkin [70], metallic zirconium has 2.6 localized electrons and 1.4 shared electrons per atom ($n_1 : n_s = 1.85$), while metallic niobium has 3.8 localized electrons and 1.2 shared electrons per atom ($n_1 : n_s = 3.17$). The statistical weight of the d^5 configuration increases materially and that of the d^0 configuration decreases as we move from zirconium to niobium.

As a result, we can expect the elements of group IV (Ti, Zr, and Hf) to transfer nonlocalized electrons to germanium atoms, the statistical weight of the d^5 levels increasing; the energy stability of these compounds should increase from titanium to hafnium as a result of the rise in principal quantum number for the d-electrons. This explains the appearance, for example, of the phase Zr_3Ge for zirconium, in which we can assume that there is strong covalent bonding between the zirconium atoms as a result of the increase in the statistical weight of the d^5 configuration. There is also a metal-rich phase for hafnium (Hf_2Ge), while titanium, for which the statistical weight and energy stability of the d^5 configuration are lowest, forms no phase of this type. Formation of similar metal-rich phases is characteristic of all metals in which the d^5 configuration can be assigned a high statistical weight (V, Nb, Ta, Cr, and Mo). In addition to transfer of electrons from transition-metal atoms to germanium atoms, strong covalent groups of germanium atoms may be formed, resulting in development of germanium-rich phases. In those cases where the metallic component can donate electrons to form a stable d^5 configuration, the donated electrons can be transferred to germanium, promoting formation of stable atomic configurations and complex structural elements of germanium atoms. One example is chromium (d^5s^1), whose s^1-electron can readily be transferred to germanium, this leading to formation of an extremely germanium-rich compound ($Cr_{11}Ge_{19}$). The probability of electron transfer to germanium atoms decreases as the number of electrons in the d shell of transition-metal atoms increases, since the statistical weight of the d^5 configuration and the acceptor capacity of the atoms rise. This leads to increased separation of the electron configurations of the metal and germanium and a decrease in their mutual overlap and in the number of shared electrons binding together the stable configurations. As the number of electrons in the d shell of transition-metal atoms rises, there is consequently a general decrease in the melting point of the corresponding germanides (see Table 2).

Examination of the magnitude of the change in the melting points of the germanide phases in the rather well-studied zirconium—germanium system shows that, in those cases where strong covalent bonds are formed between zirconium atoms (Zr_3Ge) or between germanium atoms ($ZrGe_2$), i.e., where the overlap of the electron configurations of the zirconium and germanium atoms is reduced, the melting points drop sharply to 1520-1585°C for $ZrGe_2$ and Zr_3Ge from 2240-2330°C for ZrGe and Zr_5Ge_3.

It is apparently for this reason that tungsten, in which the d^5 configuration has a high statistical weight and an extremely high energy stability, is capable neither of accepting electrons from germanium nor of donating them to it under normal thermal-excitation conditions. As a result, all attempts to produce tungsten germanides under normal conditions and at high pressures and temperatures have been unsuccessful.

Almost no research has been done on the electrophysical properties of the germanides of metals with $N_d \leq 5$, the sole exceptions being the work of Neshpor and L'vov [68] and that performed by various authors on individual germanide phases (see Table 2). From the few data available, it can be concluded that germanides are distinguished by higher electrical resistance than silicides of corresponding composition. This can be attributed to the lower energy stability of the electronic configurations in germanium atoms than in silicon atoms and the correspondingly greater electron−electron interaction. As the d-electron configuration of the metallic component becomes more stable, the resistance of the germanide drops as a result of the decrease in the proportion of unlocalized electrons.

Thus, while the resistance of $ZrGe_2$ is 120 μohm · cm, that of $NbGe_2$ ranges from 67 to 118 μohm·cm according to different data [68, 72] and that of $TaGe_2$ is 35 μohm·cm. The high resistance of $MoGe_2$ in this series (250 μohm·cm) can be attributed to transfer of the s^1 electron from the molybdenum to the germanium and formation of stable molybdenum (d^5) and germanium (sp^3) configurations, which results in a sharp decrease in the number of current carriers and a corresponding rise in resistance. This is to some extent borne out by the coefficient of thermal emf reported for $MoGe_2$ by Neshpor and L'vov [68].

Neshpor and L'vov [68] have also advanced another hypothesis to account for the nature of electrical conductivity in germanides. According to their conclusions, the fact that germanides have a higher resistance than the corresponding silicides is due to the greater atomic radius of germanium than silicon, which leads to larger relative separation of one metal atom from another in germanides than in silicides. The d levels of the metal, which scatter current carriers, are consequently constricted and become more compact. This in turn produces a decrease in current-carrier mobility from silicides to germanides.

The data on the microhardness of germanides are fragmentary but, on the whole, these compounds have a lower hardness than silicides. According to the data of Neshpor and L'vov [68], this is due to the fact that the Me−X interatomic distances are greater in germanides than in the corresponding isostructural silicon compounds. The larger Me−X interatomic distances in germanides can be attributed to the fact that the sp^3 electronic configuration is less stable in germanium than in silicon and to the larger number of shared electrons and the resultant greater openness of the lattice in germanides.

Moving on to transition metals with $N_d > 5$, it can be assumed that, in accordance with the data of Samsonov [69], the stable d^5 and d^{10} configurations combine in metallic crystals of these elements. Metals for which N_d only slightly exceeds 5 are characterized by a greater statistical weight for the d^5 configurations, while metals for which N_d is close to 10 are characterized by a higher statistical weight for the less stable d^{10} configurations.

In compounds between these metals and germanium, some of their electrons can be transferred to the germanium (and the d^5 configuration is then dominant) or electrons can be accepted from the germanium (and the d^{10} configuration is then dominant). This produces a variety of possible combinations of the bonding functions of the metal and germanium atoms and a corresponding large number of germanide phases. The presence of d^{10} configurations, which have a lower energy stability than d^5 configurations, in the germanides of these metals produces a higher concentration of shared electrons and thus a lower melting point than for germanides of metals with $N_d \lesssim 5$ (where the d^0 and d^5 configurations, which have a higher energy stability, combine). The low thermal stability of these germanides also causes many of them to have a tendency to decompose when heated (e.g., cobalt and nickel germanides).

It was pointed out above that some germanides have high temperatures of transition to the superconductive state, this being particularly true of Nb_3Ge and its derivatives produced by replacement of the germanium atoms by other elements. Since superconductivity can, in first approximation, be related to minimum capture of conduction electrons by atoms of the compound's components and to a given necessary current-carrier concentration, we can draw the following preliminary conclusion regarding the nature of superconductive compounds containing niobium and germanium. According to the data of Korsunskii and Genkin [70], 3.8-3.9 electrons are localized in metallic niobium. If we assume that roughly the same number of electrons is localized in niobium compounds, formation of ternary groupings of niobium atoms and two d^5 configurations is most probable in terms of energy; the remaining niobium electrons can then be transferred to germanium atoms, stabilizing their sp^3 configurations. Thus, two relatively stable configurations (d^5 and sp^3) are set up in the germanide Nb_3Ge; they scatter conduction electrons poorly, which causes Nb_3Ge to have a relatively high T_k (6.3-6.9°K according to different data). This value for T_k can possibly be explained by the fact that not all the excess electrons (those outside the d^5 configuration) of the niobium go to stabilize the sp^3 configuration of the germanium. There is a resultant increase in current-carrier concentration and a decrease in T_k. It is apparently for this reason that replacement of half the germanium atoms by aluminum atoms (which have an sp^2 configuration and can undergo greater stabilization than germanium atoms) accepting excess electrons from the niobium, produces a high T_k of 12.6°K ($Nb_3Ge_{0.5}Al_{0.5}$).

A similar effect evidently occurs when some of the germanium atoms are replaced by tin atoms, whose less stable (in comparison with germanium) sp configuration is capable of greater capture of niobium electrons. Compounds with the general formula $Nb_3Ge_xSn_{1-x}$ therefore have a T_k that reaches 17.6-18.0°K. More or less the same phenomenon is observed when some of the germanium atoms are replaced by gallium (sp^2): the compound $NbGe_{0.5}Ga_{0.5}$ has a T_k of 7.3°K.

The compound V_3Ge has a T_k of 6.01°K, but replacement of some of its germanium atoms by silicon atoms, which form sp^3 configurations more stable than those of germanium, raises its T_k to 14.0°K.

Lanthanide and actinide germanides are included among the transition-metal germanides. Lanthanides form germanide phases similar in composition to those of d-transition metals, as do actinides. This is due to the fact that f → d transitions are possible for both lanthanides and actinides; for example, lanthanides exhibit the transition $4f^n \rightarrow 4f^{n-1}5d$, which produces d levels. Since the 4f shell undergoes comparatively little excitation during formation of a chemical bond [73], Hund's rule [74] is applicable to it; according to this rule, the stability of the shell and the degree of binding of the electrons filling it increase with the number of possible terms. Figure 9 shows the number of possible terms, calculated from the rule $kf^n = kf^{14-n}$ (where k is the number of terms), as a function of the atomic number of the metal [75]. The maximum probability of f → d transitions corresponds to the lowest number of possible levels, except for lanthanum, whose inclusion in Fig. 9 is purely formal, since its isolated atoms have no f electrons. Lanthanides thus undergo an unusual "conversion" to d elements as a result of f → d transitions. Similar relationships are observed for actinides, the only difference being that the higher principal quantum number of the f-electrons causes the corresponding f levels and the d levels produced from them to have a higher energy stability, this being specifically responsible for the

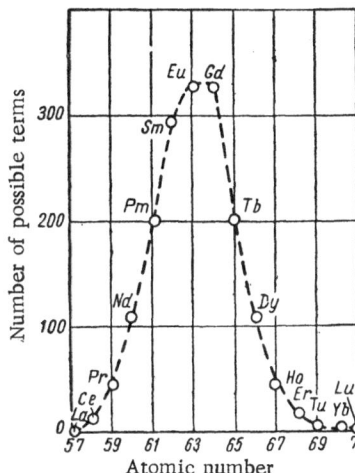

Fig. 9. Number of possible terms for lanthanides [75].

relatively high melting points of actinides in comparison with lanthanides, which exhibit equielectronic filling of the f shell.

When the electrophysical and thermal properties of rare-earth germanides having the compositions Me_5Ge_3, $MeGe$, and $MeGe_2$ (where Me is Y, La, Ce, Pr, or Nd) were investigated, it was found [467, 468] that their temperature functions of electrical resistance and thermal emf had a complex character, exhibiting inflections at definite temperatures. The type of conductivity thus changes from metallic to semiconductive; there is a decrease in the value of $d\alpha/dt$ or, in general, in the absolute magnitude of α (where α is the coefficient of absolute thermal emf and t is the temperature). Study of the coefficients of thermal expansion of rare-earth germanides established that the change in physical properties is associated with a transformation produced by a change in the electronic structure of the germanide phases. Samsonov, Paderno, and Rud' [469] attempted to relate the temperatures at which the observed changes in the properties of rare-earth germanides occur to the crystal-chemical characteristics and electronic structure of these compounds.

It can be presumed from analysis of the electrophysical properties and crystal structures of rare-earth germanides at room temperature that the Ge—Ge bonding has a covalent character and is produced by the outer (valence) s-and p-electrons of germanium, which endeavors to acquire an energetically stable sp^3 configuration. The number of bonds entered into by atoms with sp^3 configurations should increase as the phase becomes richer in germanium, i.e., as we move from Me_5Ge_3 to $MeGe_2$. It must be noted that sp^3 configurations are thermally unstable (e.g., the conversion of diamond to graphite is caused by disruption of the sp^3 configurations of the carbon atoms and formation of sp^2 configurations, which are energetically more advantageous at high temperatures). Conversely, the statistical weight of the localized d^5 levels produced in transition-metal atoms, including those of the rare-earth metals (being the result of $f \rightarrow d$ transitions in the latter), the levels on which the most densely packed structures and structures having the highest symmetry are based, is maximal at elevated temperatures, as follows from analysis of the polymorphic transformations of these metals [470].

At elevated temperatures, the energetically less stable levels (configurations) should be excited and disrupted and there should thus be an increase in the statistical weight of the most stable levels. From this standpoint, thermal excitation of rare-earth germanides can be expected to cause disruption of the sp^3 configurations of the germanium atoms, which have a comparatively low energy stability (because of the relatively high principal quantum number of the valence s- and p-electrons) and an increase in the statistical weight of the localized d levels of the metal. As a result of the localization of the shared electrons of the metal atoms and the germanium electrons freed by disruption of the sp^3 configuration, a narrow energy gap is formed between the high-density band represented by the localized d levels and the unfilled s band of the metal (the width of this gap, determined from the reciprocal temperature function, is 0.05–0.15 eV for the germanides investigated).

This change in the energy spectrum of the electrons in rare-earth germanides during thermal excitation is obviously responsible for the inversion of conductivity and for the polymorphic transformations that occur at certain temperatures. Considering that most of the structures displayed by rare-earth germanides are mutually transpositional [471], these polymorphic transformations are accompanied by only slight structural changes, being so-called displacement transformations [472].

The transformation temperatures have been determined from the inflections in the curves representing the coefficient of absolute thermal emf, resistivity, and relative elongation as functions of temperature. The following values were obtained:

Germanide	Temperature of polymorphic transformation,°C	Germanide	Temperature of polymorphic transformation,°C
Y_5Ge_3	720	Pr_5Ge_3	720
YGe	640	PrGe	490
YCe_2	460	$PrGe_2$	220
La_5Ge_3	460	Nd_5Ge_3	630
LaGe	440	NdGe	510
$LaGe_2$	250	$NdGe_2$	375
Ce_5Ge_3	150	Sm_5Ge_3	370
CeGe	580	SmGe	425
$CeGe_2$	435	$SmGe_2$	400
		$EuGe_2$	300

On the basis of the foregoing, the minimum transition temperatures (those of polymorphic transformations) should correspond to phases with a maximum relative number of bonds in which the sp^3 configurations of the germanium atoms participate. Actually, for the germanides of all rare-earth metals except cerium, the transformation occurs first in $MeGe_2$ and later (at higher temperatures) in Me_5Ge_3. It can be seen that the polymorphic-transformation temperatures range from 220 to 460°C for $MeGe_2$ phases, from 425 to 640°C for MeGe phases, and from 370 to 720°C for Me_5Ge_3 phases.

At normal temperatures, Ge—Ge bonding is effected by atoms with sp^3 configurations. By analogy with rare-earth silicides, where dislicides having a framework structure (silicon atoms with sp^3 configurations) undergo polymorphic transformations at relatively low temperatures and disilicides having structures of the AIB_2 type (silicon atoms with sp^2 configurations) do not undergo such transformations until just below their melting points, it can be assumed that the polymorphic transformations of rare-earth germanides result from disruption of the sp^3 bond hybridization and formation of stable sp^2 hybrids at elevated temperatures.

On the other hand, the change in the type of crystal structure in rare-earth germanides, like that in other properties, should also depend on a geometric factor, i.e., the atomic radii of the metals. In these germanide structures, the metal atoms are located in the voids of the germanium framework. The larger the atomic radius of the metal, the more open is the germanium framework and the lower is the strength of the Ge—Ge bonds.

As a result, disruption of sp^3 hybridization occurs first in those cases where the Ge—Ge bonds are weakest. Thus, the transition temperatures for rare-earth germanides rise monotonically as the atomic radius of the metal decreases. Cerium germanides behave somewhat anomalously. According to the available data on magnetic susceptibility, on the change in the molecular volume of germanides with increasing rare-earth atomic number, and on their x-ray spectra, it can be assumed that the valence of metals other than cerium is +3 in such compounds. The L_S x-ray absorption spectra for Ce_5Ge_3 are displaced toward the short-wave region with respect to those for CeB_6, where it has been established that the cerium is trivalent, and this indicates the presence of a substantial proportion of tetravalent cerium ions in the germanide. The fact that Ce_5Ge_3 has a much lower transition temperature than the Me_5Ge_3 germanides of other rare-earth metals can be attributed to this factor. The increase in the proportion of d-electrons (as a result of the greater probability of $f \rightarrow d$ transitions) in tetravalent cerium causes localization of the d-levels and resultant changes in properties to occur at substantially lower temperatures (about 150°C). As can be seen, this effect is more important than the decrease in metal-atom radius in the case of tetravalent cerium.

Magnetic-susceptibility and x-ray data have not shown tetravalent cerium ions to be present in $CeGe_2$ and $CeGe$. However, the somewhat anomalous decrease in molecular volume over the series of rare-earth monogermanides and digermanides nevertheless indicates that a small proportion of tetravalent cerium ions is present in addition to the trivalent ions. The dominance of the geometric factor (the decrease in the ionic radius of the cerium) is obvious in this case. The transition temperatures for $CeGe_2$ and $CeGe$ are higher than for the corresponding germanides of the adjacent rare-earth metals in the series.

According to the data of Rud',* the basic physical characteristics of rare-earth germanides have the following values:

Germanide	$n* \cdot 10^{-22}$, $1/cm^3$	$u*$, $cm/V^2 \cdot sec^2$	$\delta \cdot 10^{-23}$, $cm/V^2 \cdot sec^2$	θ, °K	$1/\rho m \theta^2$	$(\overline{u^2})^{1/2}$, Å	r ($t > t_{trans}$)	μ, eV ($t > t_{trans}$)
Y_5Ge_3	4.03	1.19	-0.51	175	0.38	0.23	1,5	0.01
YGe	2.09	3.79	-2.99	575	0.24	0.07	0,8	0.04
YGe_2	0.58	10.79	-7.00	529	0.15	0.08	1,3	0.03
La_5Ge_2	6.72	0.55	+0.20	117	0.47	0.29	1.3	0.07
$LaGe$	3.14	2.99	-2.78	445	0.36	0.08	1.2	0.02
$LaGe_2$	1.07	6.75	-4.84	425	0.23	0.09	1.2	0.02
Ce_5Ge_3	0.85	3.17	+0.85	117	0.34	0.29	1.2	0.04
$CeGe$	2.45	2.03	+1.04	464	0.18	0.08	1.0	0.09
$CeGe_2$	0.55	0.49	+2.73	455	0.10	0.08	1.5	0.02
Pr_5Ge_3	3.69	0.64	-0.15	126	0.26	0.27	0.5	0.03
$PrGe$	2.25	2.34	-0.65	463	0.13	0.08	1.1	0.02
$PrGe_2$	4.24	0.82	-0.62	466	0.08	0.08	1.1	0.03
Nd_5Ge_3	3.07	1.14	-0.39	133	0.34	0.25	1.1	0.04
$NdGe$	1.15	4.95	-2.76	492	0.17	0.07	1.0	0.03
$NdGe_2$	4.06	1.28	+0.62	500	0.12	0.08	1.1	0.02
Sm_5Ge_2	3.17	0.88	+0.14	124	0.22	0.26	1.1	0.03
$SmGe$	1.33	1.56	-0.75	207	0.46	0.22	1.2	0.02
$SmGe_2$	0.98	2.51	-0.63	200	0.33	0.24	1.0	0.03

In this table, n* and u* are the effective current-carrier concentration and mobility, $\delta = R/e\rho^2 = -(n_-u_-^2 - n_+u_+^2)$, θ is the characteristic temperature, $(\overline{u^2})^{1/2}$ is the mean square elastic-oscillation amplitude, r is the scattering parameter, and μ is the Fermi energy.

Analysis of these data enabled us to conclude that the effective current-carrier concentration for temperatures below the conductivity-inversion point $n* \approx 10^{22}$ $1/cm^3$, which is characteristic of a metallic compound (this is also confirmed by the Fermi energy, $\mu \approx 1$); at higher temperatures, μ is close to zero and r is close to 1, which demonstrates that there is a substantial proportion of ionic bonding. The maximum conductivity at this lattice-vibration amplitude is exhibited by Me_5Ge_3 germanides, as a result of the increased number of current carriers produced by enrichment of the phase in metal, while the minimum conductivity is exhibited by $MeGe_2$ digermanides. Maximum electron-phonon scattering is characteristic of Me_5Ge_3 phases.

---

* B. M. Rud', Author's abstract of dissertation [in Russian], Kiev (1968).

Rud'* measured the thermal expansion of a number of rare-earth germanides at temperatures above and below their polymorphic-transformation points:

| Germanide | Coefficient of thermal expansion | | | |
| | Below transformation point | | Above transformation point | |
	$\alpha \cdot 10^6$, deg^{-1}	Temperature range, °C	$\alpha \cdot 10^6$, deg^{-1}	Temperature range, °C
Y_5Ge_3	13.0±0 6	120-900	—	—
YGe	12.3±0.6	120-650	14.1±0.7	700-1020
YGe_2	8.9±0.4	145-475	9.4±0.5	500-800
La_5Ge_3	10.6±0.5	120-450	11.3±0.5	520-1020
LaGe	7.2±0.3	150-430	7.5±0.3	500-900
$LaGe_2$	8.0±0.4	120-300	8.9±0.4	330-1000
Ce_5Ge_3	11.4±0.5	80-150	12.7±0.6	195-750
CeGe	10.0±0.5	250-1020	—	—
$CeGe_2$	10.9±0.5	120-480	12.1±0.6	500-1000
Pr_5Ge_3	11.3±0.5	120-520	12.9±0,6	580-1000
PrGe	10.5±0.5	200-450	12.0±0.6	475-900
$PrGe_2$	10.4±0.5	100-210	11.7±0.6	240-750
Nd_5Ge_3	11.1±0.5	120-670	11.9±0.6	700-1020
NdGe	8.1±0.4	100-500	9.5±0.5	550-950
$NdGe_2$	8.2±0.4	120-300	9.6±0.5	300-820
Sm_5Ge_3	13.0±0.6	150-390	14.0±0.7	420-920
SmGe	10.0±0.5	150-450	11.2±0.5	490-1000
$SmGe_2$	10.9±0.5	124-420	11.8±0.6	470-980

As can be seen from these data, the highest coefficients of thermal expansion correspond to germanide phases with the lowest proportion of covalent Ge−Ge bonding.

The investigation of the L_S x-ray absorption spectra of certain germanides and silicides of cerium, praseodymium, and samarium (Me_5X_3, MeX, and MeX_2, where X is Si or Ge) conducted by Vainshtein et al. [453] showed that the rare-earth metals are in trivalent form in all these compounds. Analysis of the ultrastructure of the spectra of rare-earth metals in germanides and silicides indicates that the character of the interatomic interaction is similar in these compounds and that there are no atoms in anomalous valence states in such compounds as Ce_5Si_3 and Ce_5Ge_3, in whose structures x-ray diffraction studies established the presence of two types of atoms in nonequivalent crystallographic positions.

Compounds of sp-Elements with Germanium. Compounds of germanium with sp-elements (nonmetals and semimetals) can only arbitrarily be termed germanides, since most of these elements have a higher electronegativity than germanium (the electronegativities given by Pauling are 1.8 for Ge, 3.0 for N, 2.1 for P, 2.0 for As, 1.9 for Sb, 1.9 for Bi, 2.5 for S, 2.4 for Se, and 2.1 for Te). These compounds are more correctly considered to be derivatives of the corresponding semimetals and nonmetals and referred to as nitrides, arsenides, sulfides, selenides, tellurides, and so forth of germanium.

The properties of these compounds are governed by the number of sp configurations, the ability of the s-electrons to shift to p levels, the energy stability of the sp configurations, and the possibility of formation of more stable electronic configurations during their interaction.

According to the available experimental data, nonmetals and semimetals of the third group, with the exception of boron (Al, Ca, In, and Tl), do not form compounds with germanium. Since isolated atoms of all these elements have an s^2p valence-electron configuration, which can be

---

*B. M. Rud', Author's abstract of dissertation [in Russian], Kiev (1968).

converted to an sp^2 configuration by an $s \rightarrow p$ transition, while germanium has an sp^3 configuration, it can be assumed that formation of a compound is associated with a statistical possibility of transfer of one electron from the germanium to its partner. This is apparently energetically justified only in the case of boron, in which formation of an sp^3 configuration is associated with an extremely large loss of free energy and is possible in principle. Disruption of sp^3 configurations of the germanium atoms is less probable for the other elements of this group, which have their sp^2-electrons in less stable energy states.

It must be noted that compounds of germanium with boron have the compositions GeB_4 and GeB_6, i.e., organization of stable boron-atom configurations during formation of these compounds occurs principally on the basis of the boron electrons. The atomic groups formed (and probably the structural elements in the lattices of the compounds) attract a moderate number of germanium electrons for organizational purposes. The difficulty of reacting germanium with boron can also be characterized by the high activation energy for diffusion of boron in germanium, which equals 439.61 kJ/mole (105 kcal/mole).

It is possible, however, that compounds between germanium and boron are formed not by electron transfer from the germanium to the boron but by transfer of the boron electrons liberated during formation of atomic groups and corresponding stable configurations, to the germanium atoms, stabilizing their sp^3 configurations. The lack of data on the physical properties of these compounds makes it impossible for us to express a preference for one alternative or the other.

Elements of group IV (C, Si, Sn, and Pb) do not form compounds with germanium. This is probably due to the fact that carbon and silicon have a more stable equielectronic configuration than germanium, which makes electron transfer between them improbable. There are no compounds of germanium with tin and lead for almost the same reasons, but the germanium atoms are those with the configuration hard to disrupt. However, we cannot exclude the theoretical possibility of formation of compounds with tin and lead by transfer of some electrons from the germanium and a resultant increase in the stability of its sp^3 configuration.

Isolated atoms of group-V elements (N, P, As, Sb, and Bi), which have an s^2p^3 valence-electron configuration, have not been found to undergo $s \rightarrow p$ electron transitions. However, it is possible for them to transfer one electron to a partner and form an sp^3 configuration [69] on the general pattern $s^2p^3 \rightarrow sp^4 \rightarrow sp^3 \cdot p$, where p is a labile electron readily donated to atoms of another element to form compounds. In this connection, formation of compounds with these elements by germanium can be regarded as resulting from transfer of an electron from the group-V element to the germanium and stabilization of the sp^3 configurations. Both components in the compounds thus formed should have stable sp^3 configurations, with a slightly elevated electron concentration between the atoms of the germanium and the group-V element. The resistivity of the germanium nitride Ge_3N_4, 10^8 ohm·cm [77], is therefore far less than that of the silicon nitride Si_3N_4 (10^{19}–10^{20} ohm·cm), where stabilization of the silicon atoms by the nitrogen electrons is considerably more probable as a result of the higher energy stability of the sp^3 valence-electron configuration of silicon.

Similar relationships are also characteristic of the compounds of germanium with other elements in this group, but the probability of stabilization of the germanium configuration by the electrons of the group-V element decreases as we move from nitrogen to arsenic, etc. Compounds formed between germanium and arsenic therefore have a lower chemical and thermal stability and there are no compounds of germanium with antimony or bismuth.

In the case of compounds of germanium with group-VI sp-elements, whose isolated atoms have an s^2p^4 valence-electron configuration, transfer of some electrons from the germanium (and disruption of its sp^3 configuration) to the metal atoms is most probable, producing quite stable s^2p^6 configurations in the partners. This leads to formation of the compounds GeO_2, GeO, GeS_2, GeS, $GeSe_2$, $GeSe$, and $GeTe$. In connection with the aforementioned features of their electronic structure (formation of stable s^2p^6 configurations in the atoms of the group-VI elements and somewhat disrupted sp^3 configurations in the germanium atoms), all these compounds exhibit semiconductive conductivity and a broad forbidden zone, which decreases in width from compounds of germanium with oxygen to compounds of germanium with tellurium. The decrease results from the drop in energy stability and in the probability that the metal atoms will form s^2p^6 configurations as the principal quantum number of the partner increases. This explains the high current-carrier concentration detected in GeTe [78-80]. The electronic configuration of the germanium atoms is disrupted during formation of the compound, but there is a low statistical probability that the electrons liberated will be captured by the tellurium atoms (because of the low energy stability of their sp configurations). A high free-electron concentration is set up; this enables us to regard the definition of GeTe as a semimetal with an overlapping valence conduction zone [79] as correct.

The disruption of the germanium sp^3 configuration culminates in its complete breakdown in compounds with group-VII sp-elements (F, Cl, Br, and I), where, in the limiting case, all the germanium sp-electrons are transferred to the s^2p^5 configurations of the group-VII atoms to produce s^2p^6 configurations.

Proceeding from the foregoing, it must be noted that all compounds of germanium with sp-elements have chemical bonding of the covalent-ionic type.

On the basis of our consideration of the characteristics of the electronic structure, chemical bonding, and crystal structures of germanium compounds, we can thus propose the following preliminary classification of these compounds (Fig. 10):

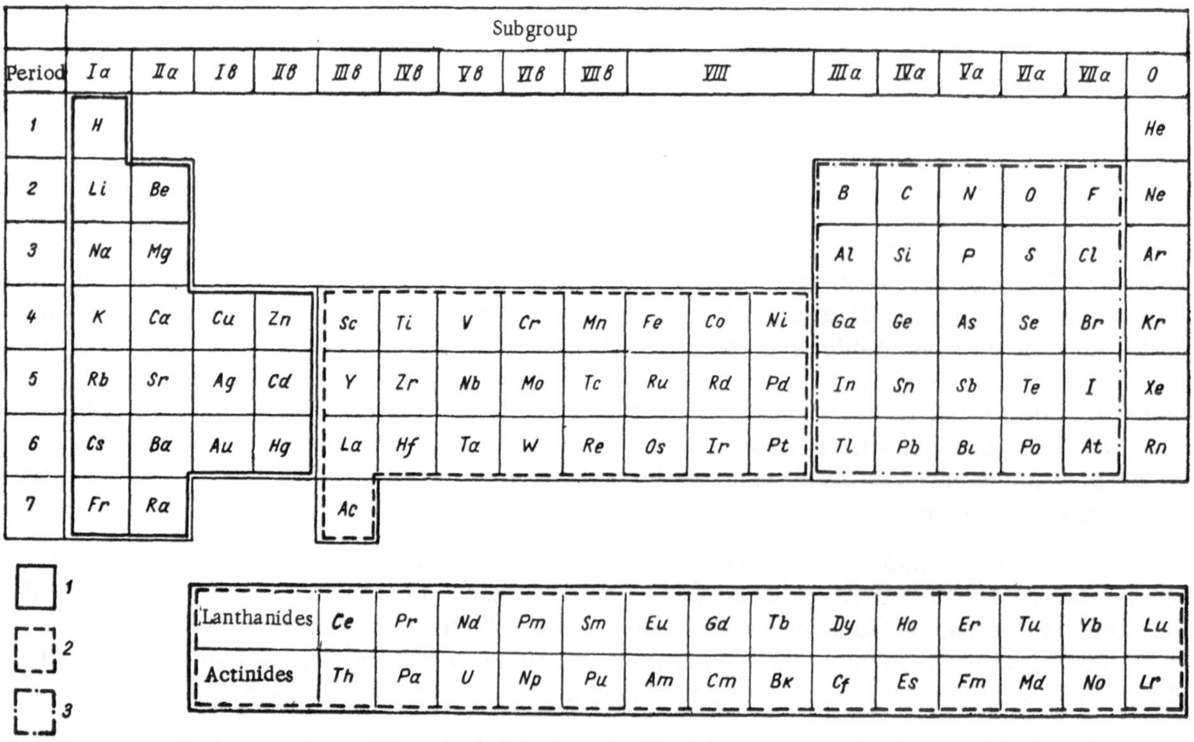

Fig. 10. Classification of compounds of germanium with the elements of the periodic table. 1) Compounds with ionic-covalent bonding; 2) compounds with covalent-metallic bonding; 3) compounds with covalent-ionic bonding.

1) compounds of germanium with s-elements (nontransition metals), in which chemical bonding is of the ionic-covalent type;

2) compounds of germanium with ds- and fds-elements (transition metals), in which the chemical bonding has a complex covalent-metallic character, with a substantial proportion of ionic bonding superimposed in a number of cases;

3) compounds of germanium with sp-elements (nonmetals and semimetals), in which bonding is principally of the covalent-ionic type.

This preliminary classification enables us to generalize from the available information on the chemical and physical properties of germanium compounds and may prove helpful in synthesizing binary and higher-order germanium compounds with predetermined properties.

CHAPTER 2

METHODS FOR PREPARATION OF GERMANIDES

METHODS

Methods for preparing germanides are similar to those for producing borides, carbides, and silicides. The following germanide-preparation methods are the best known:
1. Synthesis from components:
   a) by fusion;
   b) by sintering (or hot pressing).
2. Thermal decomposition in a vacuum.
3. Electrolysis of molten media.

Synthesis from Components

This extremely simple method was first described for high-melting compounds by Moissan [81], who employed it to produce certain metal silicides in arc furnaces. The procedure proposed by Moissan became one of the principal methods for obtaining germanides. Wallbaum [82] prepared titanium, zirconium, vanadium, niobium, and tantalum germanides by sintering powder mixtures in an argon atmosphere or a vacuum at comparatively low temperatures. As furnace technology developed, tubular graphite resistance furnaces, induction furnaces, and arc furnaces, which made it possible to achieve very high temperatures, came into use for producing germanides. Geller [83] obtained rhodium germanides by fusing the components in a helium atmosphere at $1600 \pm 200°C$. Alloys of germanium with yttrium, titanium, and zirconium were produced by arc melting. Lyashenko and Bykov [84] prepared uranium germanides and alloys by fusing high-purity components in a vacuum induction furnace.

The present authors obtained chromium and molybdenum germanides with a lattice of the β-W type by the synthesis method [85].

Nowotny [86] produced hafnium germanides by hot pressing of powders and subsequent thermal annealing, which was necessary to homogenize the alloys and relieve internal stresses. The same technique was used to prepare tantalum germanides, but tantalum hydride was used instead of the metallic powder [87].

Thermal Decomposition in a Vacuum

Hohmann [49] used this method to prepare certain alkali-metal germanides. He obtained germanides of the MeGe type by direct synthesis and then heated these compounds in a high vacuum to produce compounds of the $MeGe_4$ type. Potassium, rubidium, and cesium germanides were obtained in the same manner.

Electrolysis of Molten Media

Barbier-Andrieux [88, 89] isolated molybdenum germanides from alloys prepared by electrolysis of sodium or lithium metaborate, germanium dioxide, and molybdic anhydride. He used the same technique to prepare alloys of germanium with iron, cobalt, manganese, and nickel.

ALKALI-METAL GERMANIDES

Lithium Germanides. Pell [59] established the existence of two compounds: $Li_{4n}Ge_n$ (probably Li_4Ge) and $Li_{3n}Ge_n$ (probably Li_6Ge_2).

Gladyshevskii and Kripyakevich [60], who investigated alloys containing 14, 17, 20, 23, and 25 at-% Ge, showed that the compound Li_4Ge must be assigned the formula $Li_{15}Ge_4$. The alloys were prepared by fusing lithium (98.88% Li) and germanium (99.99% Ge) in iron crucibles in an argon atmosphere or in quartz crucibles under an LiCl flux, in an electric furnace. The resultant alloys were brittle, exhibited metallic fracture, and readily oxidized in air.

Sodium Germanides. Dennis and Skow [90] produced the compound NaGe by heating germanium and an equimolar amount of sodium to 1000°C in a closed steel crucible. When the crucible was opened, they found a dense, hard, readily combustible product of uniform composition.

Hohmann [49] produced this compound by the same technique, the only difference being that he used a three- or four-fold excess of the alkali metal instead of an equimolar amount. Finely pulverized germanium (0.5 g) was reacted with sodium (0.7-0.8 g) at 650-700°C in an argon atmosphere in corundum crucibles placed in hermetically sealed steel bombs. After two days, the resultant product was slowly cooled to room temperature and the excess sodium was driven off. The product obtained was loose and consisted of small, distinct acicular crystals with a metallic luster.

Potassium Germanides. The compound KGe was produced by synthesis from its components at 600°C, using the technique described by Hohmann [49] for NaGe. After removal of the excess potassium, a loose, dark-colored product containing no distinct crystals was obtained.

Johnson [3] prepared this compound in the form of crystals of indeterminate shape by heating powdered germanium with small chunks of potassium at 1000°C in a steel crucible lined with scale-resistant steel.

The compound KGe_4 was produced by thermal decomposition of KGe under high vacuum [49]. For this purpose, 200-300 mg of KGe was loaded into a fired corundum crucible and very slowly heated in a glass tube about 25 cm long, two-thirds of which was placed in an electric furnace. The decomposition temperature of the KGe was 400-420°C.

Rubidium Germanides. The compound RbGe was obtained by direct synthesis at 600°C and subsequent evaporation of the excess rubidium at 200°C. The compound had the form of crystals with a bronze color.

The compound $RbGe_4$ was obtained by thermal decomposition of RbGe at 390-410°C under high vacuum (in the same manner as for the Ge−Na and Ge−K systems [49, 64]).

Cesium Germanides. The compound CsGe was obtained by direct synthesis at 600°C and subsequent evaporation of the excess cesium at 180°C. It had the form of attractive black crystals.

The compound $CsGe_4$ was produced by thermal decomposition of CsGe at 390-410°C under high vacuum (in the same manner as for the Ge−Na and Ge−K systems [49, 64]).

GERMANIDES OF COPPER-SUBGROUP METALS

Copper Germanides. Microscopic and x-ray analyses of germanium−copper alloys prepared by fusion of the electrolytic metals in evacuated ampules established the existence of the chemical compounds Cu_5Ge (18.59% Ge) and Cu_3Ge (27.58% Ge) [92, 94, 97, 98].

ALKALINE-EARTH GERMANIDES

Magnesium Germanides. Klemm and Westlinning [99] prepared the compound Mg_2Ge by fusion of the pure metals in a corundum crucible placed in a steel bomb. Argon was employed as the protective atmosphere.

Brauer and Tisler [100] used a graphite crucible to produce Mg_2Ge alloys. The crucible and charge were placed in a hermetically sealed argon-filled steel bomb and heated at 1200-1230°C in an electric furnace for 30 min. The resultant product was dense and had a high hardness. Fresh fractures showed a finely crystalline structure with a silvery gray color.

Kroemer et at. [36] produced high-quality Mg_2Ge monocrystals with a predetermined content of alloying impurities (As, Sb, B, and U). The compound was prepared by fusing stoichiometric amounts of magnesium and germanium in a graphite crucible placed in a hermetically sealed tantalum bomb. Bridgman's method was used to obtain monocrystalline specimens.

Calcium Germanides. The compounds $Ca_{33}Ge$ (5.20% Ge), Ca_7Ge (20.56% Ge), Ca_2Ge (47.53% Ge), CaGe (64.43% Ge), and $CaGe_2$ (78.36% Ge) were produced by fusing metallic germanium and calcium at 1150-1200°C in an argon atmosphere and slowly cooling the resultant product [55, 101-104].

The technique used to prepare these compounds was similar to that employed for Ca_2Si [55], its main features consisting of the following. A corundum crucible containing the charge was placed in a hermetically sealed iron crucible, which was evacuated, filled with argon, and heated in a furnace at 1150-1200°C for 1 h. It was then slowly cooled to 860°C and subjected to prolonged isothermal holding at this temperature, which, in the author's opinion, facilitated completion of the peritectic reaction. At the end of the holding period, the crucible was slowly cooled to 700°C and then to room temperature.

A charge consisting of 3.1 g of Ge and 4.5 g of Ca was specifically used to prepare the compound Ca_2Ge.

The compound CaGe was first produced by Royen and Schwarz [102]. It was prepared by heating of calcium hydride and germanium powders in an electric furnace at 950°C, the reaction taking the form $CaH_2 + Ge = CaGe + H_2$. The reaction was judged to have gone to completion when hydrogen evolution ceased.

This same compound was later prepared by Eckerlin et al. [101]. A charge consisting of 4.5 g of Ca and 6.2 g of Ge was fused at 1200°C, very slowly cooled to 930°C, and then abruptly cooled to room temperature. The cooled melt contained irregular silvery gray chunks of alloys, which consisted partially of very finely crystalline material and partially of flat monocrystals up to 0.5 mm long.

Alloys with the compositions $Ca_{33}Ge$ and Ca_7Ge were also prepared by the method described above [55] and consisted of cubic and acicular crystals [103].

TRANSITION-METAL GERMANIDES

Titanium Germanides. Pietrokowsky and Duwez [105, 106] produced the compound Ti_5Ge_3 by arc-fusion in a water-cooled copper crucible and subsequent homogenization in evacuated tubes at 1038°C for 3 days.

Ageev and Samsonov [107] obtaine the compound TiGe by sintering a pressed powder mixture at 900°C for 20 h. The initial components were titanium iodide (99.7% pure) and germanium (99.9% pure). The procedure for preparing TiGe was basically similar to that for preparing the titanium silicide TiSi, its main features consisting of the following. A charge containing mixed metallic powders of the requisite composition was pressed into cylindrical specimens

20 mm in diameter and 100 mm long, which were then sealed into evacuated quartz ampules and sintered at different temperatures and holding times. The structure and lattice parameters of the TiGe were determined in monocrystals taken from shrink holes in an ingot prepared by fusing the sintered specimens in a thorium oxide crucible. Fusion was carried out by induction heating from a high-frequency generator. For this purpose, the crucible and charge were placed in an evacuated quartz container fitted with an internal graphite heating element, whose size and shape ensured controlled crystallization of the melt and formation of shrink holes.

Zirconium Germanides. Wallbaum [82] produced the compound $ZrGe_2$ by a powder-metallurgical method and determined its structure.

Parthe and Norton [87] prepared the compound Zr_5Ge_3 by hot pressing of thoroughly dried zirconium hydride and germanium powders in graphite pressforms. The compound was produced at a temperature of 1200°C and a holding time of 2 min.

According to our data, Zr_5Ge_3 can be obtained by cold pressing of metallic powders with subsequent sintering and by hot pressing at 900-1750°C in an argon atmosphere for 3-5 min.

Neshpor and L'vov [68] prepared compact specimens of a zirconium germanide with the composition $ZrGe_2$ and investigated its physical properties. The specimens were obtained by sintering metallic zirconium and germanium powders in an argon atmosphere at 1300°C with subsequent sintering of the resultant germanide by hot pressing.

Hafnium Germanides. Parthe [108] obtained the compound Hf_5Ge_3 by hot pressing of a mixture of hafnium hydride and metallic germanium with 3 at-% C added to stabilize the compound. Sintering was carried out at a temperature of 1200°C and a holding time of 2 min.

Nowotny et al. [86] comfirmed the existence of Hf_5Ge_3 and $HfGe_2$ and established that of Hf_2Ge and $HfGe$. The starting materials were sponge hafnium, which contained 2.2% Zr, 0.01% Mg, 0.01% Cl, 0.007% Fe, 0.004% Si, and other impurities, and high-purity powdered germanium. The sponge hafnium was preliminarily hydrided in flowing purified hydrogen. The compounds were obtained by hot-pressing the mixture of hafnium hydride and germanium at 1000°C in an argon atmosphere and sintering it at 1000°C for 4 h, with subsequent homogenization at 1700°C.

Vanadium Germanides. The germanium-vanadium system has been found to contain four chemical compounds: V_3Ge (32.20% Ge), V_5Ge_3 (46.09% Ge), V_3Ge_2 (48.71% Ge), and VGe_2 (74.02% Ge) [82, 109-111].

Wallbaum [82] employed powder-metallurgical methods to produce V_3Ge and described its structure.

Hardy and Hulm [109, 110] obtained this compound by sintering pressed metallic vanadium and germanium powders at 1000°C in a purified helium atmosphere for several days.

Holleck et al. [111] confirmed the existence of the chemical compound V_3Ge and established that of the new compounds V_5Ge_3, V_3Ge_2, and VGe_2. Specimens of alloys with different compositions were prepared by cold pressing of powdered vanadium (produced by hydriding metallic vanadium) and powdered germanium and hot pressing at the same temperatures for 8-15 h in an argon atmosphere, with subsequent homogenization at 900-1250°C.

The compound V_3Ge_2 was prepared by pressing and subsequent sintering.

Niobium Germanides. Wallbaum [82] and Hardy [109, 110], who investigated specimens prepared by a powder-metallurgical method, i.e., sintering of pressed metallic powders at 1000°C in a purified helium atmosphere for several hours, and by fusion of pressed rods in an arc furnace under argon, discovered the existence of the compounds Nb_2Ge and $NbGe_2$.

Carpenter and Searcy [113] prepared the compounds Nb_3Ge, α-Nb_5Ge_3, β-Nb_5Ge_3, and $NbGe_2$. The starting materials were powdered niobium (99.4% pure) and germanium (99.9% pure), the latter being preliminarily heated in a vacuum to remove any oxides. The niobium contained small amounts of carbon and tantalum as its main impurities.

The latter three compounds were produced by sintering mixtures of the initial metallic powders in graphite or tungsten crucibles in a vacuum at 1400°C for 30 min or 1600°C for 1 h.

The compound Nb_3Ge was obtained by heating $NbGe_{0.54}$ and Nb in a tungsten crucible at 1700°C for more than 3 h.

Carpenter also produced this same compound by a different method [114]. A mixture of the metallic powders was heated in molybdenum, tantalum, or niobium crucibles in a vacuum at 1750°C for 6 h. The resultant specimens were then pulverized and resintered in a vacuum in order to remove the excess germanium and achieve equilibrium. This procedure was continued until pure monophasic specimens of Nb_3Ge were obtained.

According to our data, the compound Nb_5Ge_3 can be obtained by hot pressing of metallic powders in metallic pressforms in an argon atmosphere at 1400-1600°C for 3-5 min.

Neshpor and L'vov [68] produced compact specimens of $NbGe_2$ by sintering niobium and germanium powders in an argon atmosphere at 1300°C and then sintering the resultant powdered germanide by hot pressing. According to our data, this compound can also be obtained by sintering the metallic powders by hot pressing at 1100-1250°C for 3-5 min.

Tantalum Germanides. Wallbaum [82] employed powder-metallurgical methods to prepare the compound $TaGe_2$ and determined its structure.

Parthe and Norton [87] produced the compound Ta_5Ge_3 by hot pressing of tantalum hydride and germanium powders with 5 at-% C added at 1200°C for 2 min.

According to our data, this compound can also be obtained by hot pressing of metallic germanium and tantalum powders in graphite pressforms in an argon atmosphere at a temperature of 1250-1750°C for 3-5 min.

Chromium Germanides. Wallbaum [82] discovered the germanides Cr_3Ge, Cr_3Ge_2, and $CrGe$ in specimens prepared by powder-metallurgical methods and determined their structures.

The present authors [85, 116] obtained Cr_3Ge by fusion of high-purity germanium and chromium powders in quartz crucibles. For this purpose, thoroughly dried germanium and chromium powders with a grain size of 50 μ were made up into a stoichiometric charge and thoroughly mixed for 12 h. The charge was fused in an electric resistance furnace, which was preliminarily evacuated and then filled with purified argon to a residual pressure of about 20 kN/m^2 (0.2 atm). Fusion was carried out at 1520-1540°C for 1-2 min. The temperature was then abruptly reduced to 500°C and the specimens were quenched in water from this temperature. The specimens were brittle and exhibited silvery, lustrous, finely crystalline fracture.

Compact specimens were prepared from the resultant compound in the following manner. The Cr_3Ge specimens were ground in an agate mortar and sieved through a screen with 50-μ apertures. Cylindrical specimens 8 mm in diameter and 10-15 mm long were made up from the germanide powder thus obtained and sintered in a mechanical lever-weight press heated by passing current through a graphite matrix with a diameter of 30 mm. The specimens were sintered at 1450°C for 3-5 min in an argon atmosphere; their density amounted to 97-98% of the theoretical figure.

Parthe and Norton [87] obtained the compound Cr_5Ge_3 by sintering and hot pressing of chromium and germanium powders in graphite pressforms at 1200°C for 2 min.

Molybdenum Germanides. Methods for the production of Mo_3Ge have been studied in rather great detail. This compound was first obtained by Searcy et al. [221], who heated a mixture of molybdenum and germanium powders with an atomic ratio of about 3:1, at 1000°C.

Hardy and Hulm [109, 110] obtained this compound and determined the temperature at which it enters the superconductive state. The Mo_3Ge was prepared by sintering pressed metallic powders in rod form at 1000°C in purified helium for several hours.

Bondarev [116] prepared compact Mo_3Ge specimens and studied certain of their physical properties (see Table 2). The compound was produced by the technique described in a previous article [85], whose main points consisted of the following. Thoroughly dried metallic powders were made up into a stoichiometric charge, thoroughly mixed for 12 h, and pressed into rods 8 mm in diameter and 12-14 mm long under a pressure of 1372-1470 MN/m^2 (14-15 T/cm^2).

The pressed rods were sintered on a molybdenum support in a vacuum resistance furnace, which was preliminarily evacuated and filled with argon to an overpressure of 19.6 kN/m^2 (0.2 atm). Sintering was carried out at 1700-1800°C for 4 h. At the end of the holding period, the specimens were cooled in 150-deg steps and quenched in water from 600°C. The resultant specimens were brittle and exhibited silvery, lustrous, finely crystalline fracture.

The compact specimens were prepared by the following procedure. The germanide specimens were ground in an agate mortar and sieved through a screen with 50-μ apertures. Specimens 8 mm in diameter and 10-15 mm long were sintered from the resultant powder in graphite pressforms in a mechanical lever-weight press heated by passing a current through a graphite matrix 30 mm in diameter. The Mo_3Ge specimens were sintered at 1650-1740°C for 3-5 min; their density was 94-95% of the theoretical figure.

Searcy and Peavler [118] obtained the compound Mo_3Ge_2, as well as Mo_2Ge_3, α-$MoGe_2$, and Mo_3Ge by heating mixtures of germanium and molybdenum powders with a grain size of 70 μ in graphite crucibles at 980°C for 5-6 h. Sintering was carried out in a resistance furnace in an argon atmosphere.

These compounds, together with β-$MoGe_2$, were also obtained by heating the same mixtures in an induction furnace in a vacuum at 1350°C. The molybdenum powder lost considerable mass in both cases, while the germanium powder exhibited a loss of several milligrams.

In order to reduce the losses in subsequent experiments, crushed chunks of germanium were used instead of the powder and the crucible was covered with a lid having drilled holes 0.8 mm in diameter. The germanide specimens produced at 980°C were cooled at a rate of 100 deg/h, while those produced at 1350°C were cooled to 700°C over a period of 3-4 min and then permitted to cool with the furnace.

Neshpor and L'vov [68] obtained compact $MoGe_2$ specimens by sintering molybdenum and germanium powders in an argon atmosphere at 1300°C and then sintering the resultant germanide powder by hot pressing. The molybdenum germanide specimens produced contained two phases, which corresponded to the two crystalline forms of $MoGe_2$ in composition.

Manganese Germanides. The compounds $Mn_{3.25}Ge$ (28.91% Ge), Mn_5Ge_2 (34.58% Ge), Mn_5Ge_3 (44.22% Ge), and Mn_3Ge_2 (46.84% Ge) were discovered by Zwicker et al. [119] in investigating alloys prepared by fusing electrolytic manganese and distilled germanium. Charges having different compositions and masses of 1-5 g were fused in an electrolytic-hydrogen atmosphere and then subjected to prolonged high-temperature annealing. In some cases, prolonged vacuum annealing of small specimens led to severe evaporation. The annealing was therefore conducted in two stages, i.e., the specimens were first homogenized below the solidus in a hydrogen atmosphere and then annealed in a vacuum at lower temperatures.

Iron Germanides. Shtol'ts and Gel'd [122-125], who studied the germanium−iron system, established the existence of four chemical compounds ($Fe_{3.25}Ge$, Fe_3Ge, Fe_4Ge_3, and FeGe) in addition to the two that had previously been discovered (Fe_2Ge and $FeGe_2$).

The alloys were prepared by fusing reduced iron and monocrystalline germanium in sealed quartz ampules with subsequent prolonged homogenization. Specifically, an alloy having the composition $Fe_{3.25}Ge$ was vacuum-annealed for 100 h at 1000°C. In contrast to the brittle compounds Fe_2Ge and $FeGe_2$, the $Fe_{3.25}Ge$ obtained had a rather high viscosity and hardness, which made it very difficult to pulverize it. In order to make an x-ray diffraction analysis, the pulverized alloy was therefore given an additional annealing at 1000°C for 3 h, to relieve internal stresses. Alloys having the compositions Fe_3Ge, Fe_4Ge_3, and FeGe were also subjected to prolonged homogenization annealing.

According to Ohoyama's data [127], the compound FeGe was detected in specidents prepared by heating a mixture of the components in sealed quartz ampules at 1300°C for 2 h, annealing, and quenching from a temperature below 700°C.

Lecocq and Michel [128] prepared alloys for thermomagnetic investigations by the cermet method. Germanium and iron powders were sintered for 7 h at 800°C with subsequent quenching in water.

Cobalt Germanides. Laves and Wallbaum [129] were the first to report the existence of chemical compounds with the compositions Co_2Ge and $CoGe_2$.

Pfisterer and Schubert [130] confirmed the existence of the aforementioned germanides and established that of the new compounds CoGe and Co_2Ge_3. Metallic cobalt (98% pure) preliminarily treated in flowing hydrogen at 1200°C for 0.5 h and germanium containing traces of magnesium, copper, and chromium (less than 0.001% total) were used to prepare the alloys. Alloys with a mass of 5-8 g were melted in a Tamman furnace in an electrolytic-hydrogen atmosphere and then annealed at high temperatures.

The compounds CoGe and Co_5Ge_8 were produced by Bhan and Schubert [131], who pointed out that they were prepared by fusion in evacuated quartz ampules in an arc furnace in an argon atmosphere. Specifically, the CoGe was obtained in the following manner. A charge having a mass of 2.5 g and the composition CoGe was fused in an evacuated quartz ampule and then cooled with the furnace from 1000 to 800°C over a four-hour period. Long-term annealing was carried out at this temperature for 12 h.

Stolz and Schubert [132] reported the compound Co_5Ge_7.

Barbier-Andrieux [88, 133] produced an alloy with the composition $Co_{2.4}Ge$ by electrolysis of molten sodium germanate, cobalt oxide, and sodium fluoride at 1050°C.

Nickel Germanides. Ruttewit and Masing [135] were the first to report the existence of the germanides Ni_3Ge (29.19% Ge), Ni_2Ge (38.21% Ge), and NiGe (55.29% Ge).

The compounds Ni_3Ge and NiGe were later obtained [134] by fusion of the components in a hydrogen−nitrogen atmosphere in a Tamman furnace or in evacuated quartz ampules in an oxygen-burner flame. Beryllium oxide or porcelain was used as the crucible material.

Castellitz [137] obtained an alloy with the composition $Ni_{1.70}Ge$. A compound was prepared by fusion or sintering of metallic nickel and germanium (preliminarily vacuum-remelted) in magnesium oxide or aluminum oxide crucibles, followed by homogenization for 2-3 h at a temperature below the solidus and very slow cooling to room temperature.

Ruthenium Germanides. The compounds RuGe and Ru_2Ge_3 were produced by fusion of the initial metals in an arc furnace with a nonreactive tungsten electrode in an argon atmosphere and subsequent annealing in evacuated quartz tubes [138].

Rhodium Germanides. Geller [83] established the existence of the chemical compounds Rh_2Ge (26.08% Ge), Rh_5Ge_3 (29.73% Ge), and RhGe (41.36% Ge) in the germanium−rhodium system. The alloys were prepared by fusing mixtures of high-purity germanium and rhodium powders in graphite crucibles in an induction furnace at 1600 ± 200°C, followed by cooling with the furnace. Both fusion and cooling were carried out in a purified-helium atmosphere.

Palladium Germanides. Pfisterer and Schubert [134], who investigated alloys in the germanium−palladium system, established the existence of the compounds Pd_4Ge, Pd_5Ge_2, Pd_2Ge, and PdGe. Alloys with a mass of 0.2 g were prepared by fusion of metallic palladium and germanium powders in beryllium oxide (or porcelain) crucibles in a Tamman furnace in a nitrogen atmosphere and subsequent homogenization in evacuated quartz tubes for 2 days.

Anderko and Schubert [139] reported production of the compound Pd_2Ge and studied its structure. For this purpose, a mixture of pure metallic palladium and germanium powders was fused in evacuated quartz ampules or sintered in corundum crucibles in a nitrogen atmosphere in a Tamman furnace. Since the reaction of the metallic powders and formation of Pd_2Ge involved a large exothermic effect, the furnace temperature was held below the melting points of the initial components.

Osmium Germanides. The existence of the compound $OsGe_2$ was first reported by Wallbaum [82], who studied specimens prepared by a powder-metallurgical technique.

Weitz et al. [140] produced the compound $OsGe_2$ and determined its structure. It was prepared by fusing the pure elements in a corundum crucible placed in an evacuated quartz tube. The structure of the $OsGe_2$ was determined with monocrystals pulled from the melt.

Iridium Germanides. The compounds IrGe and Ir_3Ge_7 were produced by fusing high-purity germanium and iridium (the latter with a purity of 99.9%) in evacuated quartz tubes, with subsequent very slow cooling with the furnace [134]. The resultant compounds were homogeneous.

The compounds Ir_4Ge_5 and $IrGe_4$ were produced by arc-fusion of the high-purity components in an argon atmosphere [131].

Platinum Germanides. The procedure used to prepare platinum germanides was basically analogous to that employed for nickel, palladium, and iridium germanides. Thus, the compound PtGe was produced by fusion of the pure elements in evacuated quartz tubes or beryllium oxide crucibles in a Tamman furnace in a nitrogen atmosphere [134]. Alloys having the compositions Pt_2Ge and Pt_2Ge_3 were also prepared by this method.

Anderko and Schubert [139] obtained the compound Pt_2Ge by the same method, but used corundum crucibles instead of beryllium oxide crucibles.

Bhan and Schubert [131] prepared alloys found to contain the compounds Pt_3Ge, Pt_3Ge_2, Pt_2Ge_3, and $PtGe_2$ by heating the initial components (with a total mass of 1 g) in evacuated quartz ampules or by fusing them in an argon atmosphere in an arc furnace and then homogenizing them. Specifically, the compound Pt_3Ge was obtained by fusion of the components at 1000°C and subsequent homogenization at 920°C for 24 h.

Germanides of Rare-Earth Metals and Actinides. The data on procedures for producing germanides of rare-earth elements and actinides are extremely limited. Analyzing them, it can be noted that the known germanides were produced by two methods: fusion in quartz tubes in Tamman funaces or arc-fusion in an inert atmosphere.

Thus, an alloy with the composition Y_5Ge_3 (32.88% Ge) was obtained by arc-fusion in a helium atmosphere and then thrice remelted in order to even out its composition [141].

The compounds Gd_5Ge_3 (21.69% Ge) and Dy_5Ge_3 (21.14% Ge) were also obtained by arc-fusion [142].

The compounds La_5Ge_3 (23.87% Ge), Ce_5Ge_3 (23.71% Ge), and Sc_5Ge_3 (49.20% Ge) were produced by fusion of pressed rods consisting of powdered germanium and metallic rodlets of the corresponding elements in an inert atmosphere in an induction furnace.

The monogermanides ScGe and YGe [199] were obtained by induction fusion of pressed rods in boron nitride crucibles in a purified-argon atmosphere. The fused alloys were homogenized in sealed quartz ampules evacuated to less than 13.33 mN/m^2 (10^{-4} mm Hg) at 1000-1200°C for no more than 4 h. Longer holding times caused the specimens to react with the quartz to form oxides. Thus, the reaction between ScGe and quartz (at 1200°C with a holding time of 20 h) can be described by the equation

$$ScGe + SiO_2 = Sc_2O_3 + \text{mixture } (Ge-Si) + ScGe_2.$$

Gladyshevskii [145, 146] produced digermanides of the rare-earth metals La, Ce, Nd, Sm, Eu, Tb, Dy, Ho, Er, Tu, Yb, and Lu and studied their structures. The alloys were obtained by fusion of rare-earth metals with 99.9% germanium in evacuated quartz ampules in a Tamman furnace or by arc-fusion with a tungsten electrode in a water-cooled copper crucible in a helium atmosphere.

Lyutaya and Goncharuk [147] prepared the compounds La_5Ge_3, $LaGe_2$, and LaGe by fusion of the pure metals (99.2% La and 99.9% Ge) in an arc furnace with a tungsten electrode in an argon atmosphere and subsequent triple remelting to even out their composition, as well as in a quartz reaction vessel in flowing pure, dry argon in a resistance furnace. Lanthanum digermanide was also produced by germaniothermal reduction of pure lanthanum oxide in a vacuum furnace with a molybdenum heating element at 1500°C for 1 h with an initial furnace pressure of 133.3-13.33 mN/m^2 (10^{-3}-10^{-4} mm Hg). The charge was prepared on the basis of the presumed reaction

$$La_2O_3 + 7Ge = 2LaGe_2 + 3GeO.$$

The germaniothermal method was later investigated in greater detail by Kosolapova and Lynchak [483, 484], who showed that the corresponding reactions pass through a stage in which rare-earth germanates are formed; at temperatures above 1200-1300°C, these are reduced to germanides. As a result of the severe evaporation of germanium at furnace pressures of the order of 10^{-3} mm Hg and temperatures of the order of 1400-1500°C, it is impossible to obtain digermanides: only germanium-poor phases with the composition Me_5Ge_3 are formed. The furnace pressure must be of the order of 10^{-1}-10^{-2} mm Hg for $MeGe_2$ digermanides to be produced, the optimum process temperature being 1450°C.

Thorium Germanides. Alloys having the compositions Th_3Ge (9.44% Ge), Th_3Ge_2 (17.26% Ge), ThGe (23.83% Ge), α-$ThGe_2$ and β-$ThGe_2$ (38.48% Ge), and $ThGe_3$ (48.42% Ge) were prepared at temperatures between 1000 and 1500°C by powder-metallurgical methods [148].

The compound $ThGe_2$ and the new compound $Th_{0.9}Ge_2$ were prepared by Brown [149], who employed the following method. Thorium (99.4% pure) and germanium (99.999% pure) were refused in different ratios in a water-cooled copper crucible in an argon atmosphere in an arc furnace with a nonreactive electrode. Each ingot was turned and remelted several times to homogenize it. After fusion, 10 g of the alloy, which had a Ge content of about 60 at-%, was sealed into an evacuated quartz ampule with 40 g of bismuth. The ampule was heated at 700°C for 48 h, with frequent agitation. At the end of the holding period, the product was cooled, broken into chunks, and transferred to an alumina crucible. The crucible and its contents were again sealed into an evacuated quartz ampule and held at 700 ± 10°C for one month. The resultant $Th_{0.9}Ge_2$ crystals, which ranged in size from 50 to 200 μ, and the free germanium were isolated by dissolving the bismuth base in 50% nitric acid.

The compound $ThGe_2$ was obtained in similar fashion, the only difference being that the crystals were grown at a temperature no higher than 600°C.

Using the previous method [149], Brown and Norreys [150] produced a new compound, $ThGe_{1.5}$ (31.93% Ge), in addition to the aforementioned thorium germanides. This compound was prepared by heating arc-fused thorium alloys in liquid bismuth or lead at a temperature below 600°C for 100 h and then isolating it by etching out the bismuth base with nitric acid.

Uranium Germanides. Alloys in which the compounds U_7Ge (4.17% Ge), U_5Ge_3 (15.47% Ge), U_3Ge_4 (28.90% Ge), UGe_2 (37.88% Ge), and UGe_3 (47.78% Ge) were found were prepared by fusing 99.86% pure uranium and 99.99% pure germanium [84]. Fusion was carried out in beryllium oxide crucibles in a vacuum induction furnace at a pressure of 133.3–1.333 mN/m^2 (10^{-3}–10^{-5} mm Hg). All the alloys were then homogenized at 900°C for 150 h. The alloys containing 40–75 at-% Ge were additionally annealed at 1350°C for 25 h, with subsequent holding at 1000°C for 100 h. They were quenched in oil for study of the high-temperature regions of the germanium–uranium system.

Methods for producing germanides of elements in the boron, carbon, and nitrogen subgroups will be considered below.

CHAPTER 3

GERMANIDES OF METALS IN GROUPS I AND II

ALKALI-METAL GERMANIDES

Lithium Germanides. In studying alloys of lithium with germanium, Pell [59] detected two compounds in them: $Li_{4n}Ge_n$ (probably Li_4Ge) with a melting point of 750°C and $Li_{3n}Ge_n$ (probably $LiGe_2$) with a melting point of 800°C. A eutectic that melts at 525°C is formed in this system at a lithium content of 49 at-%.

Gladyshevskii and Kripyakevich [60], who investigated the compound Li_4Ge, showed that it must be assigned the formula $Li_{15}Ge_4$. According to their data, it belongs to the T_d^6–J43d space group, with the atomic positions

$$12\ Li^{(1)}\ in\ 12\ (a);$$
$$48\ Li^{(2)}\ in\ 48\ (e);\ x = 0.12,\ y = 0.16,\ z = 0.96;$$
$$16\ Ge\ in\ 16\ (c);\ x = 0.208.$$

The structure of $Li_{15}Ge_4$ consists of very dense packing of atoms of different sizes. It can be represented as a packing of polyhedra $(GeLi_3^{(1)}\ Li_9^{(2)})$, resembling the Cr_3Si structure in this respect.

Johnson et al. [151] clarified the composition and structure of $Li_{15}Ge_4$. According to their data, it has a face-centered cubic lattice of the $Cu_{15}Si_4$ ($D8_6$) type with the constant a = 1.078 nm, which is in complete agreement with the data of Gladyshevskii and Kripyakevich [60].

Gladyshevskii et al. [152] later established the existence of $Li_{22}Ge_5$ in addition to $Li_{15}Ge_4$.

In order to study lithium-germanium alloys containing 15, 17, 19, 20, 21, 22, 23, 25, and 28.6 at-% Ge, lithium (98.88%) and germanium (99.9%) were fused in iron crucibles in an argon atmosphere in an electric furnace. X-ray diffraction analysis of the unannealed alloys and of alloys annealed at 200°C for 3 days showed the presence of four chemical compounds. The first two (in alloys containing 15, 17, and 28.6 at-%) were isostructural with $Li_{22}Si_5$ and Li_2Si, the third (in alloys containing 19 and 20 at-%) was isostructural with $Li_{15}Ge_4$, and the fourth (in alloys containing 21–26 at-%) had a structure derived from the α-Fe type but differing somewhat from that of the compound of similar composition in the Li–Si system.

The compound $Li_{22}Ge_5$ has a cubic lattice of the $Li_{22}Pb_5$ type (F 23 – T^2) with the constant a = 1.886 nm.

Fedorova and Molochko [436] employed thermal and microstructural analyses to study the phase diagram of the lithium–germanium system, which is shown in Fig. 11. It was investigated between 0–4.5 and 42–100 at-% Ge. As can be seen from Fig. 11, the liquidus lies somewhat below that calculated by Pell [59] at 50–100 at-% Ge. Two eutectics are formed in the system: one in the vicinity of 50 at-% Ge (with a melting point of 528°C) and the other in the region below 0.2 at-% Ge (with a melting point of 184°C).

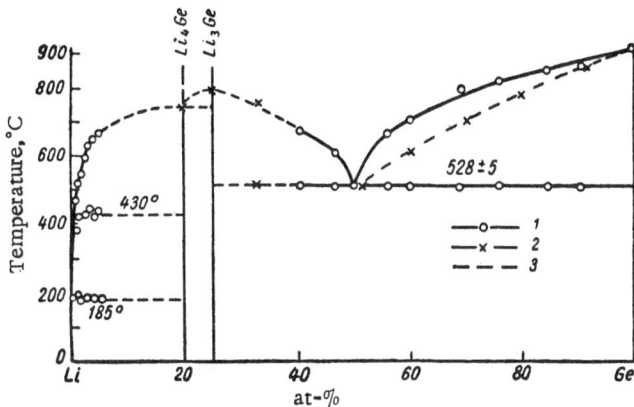

Fig. 11. Phase diagram of the lithium−germanium
system [436]. 1) Liquidus, according to Pell [59];
2) points determined by Pell [59]; 3) proposed form
of diagram in uninvestigated region.

According to the data of these authors, the compound $Li_{3n}Ge_n$ takes the form of long acicular crystals with a steel-gray color and a metallic luster, their fracture surfaces rapidly being covered by a grayish-violet film.

Only very small amounts of lithium dissolve in solid germanium and this solubility has
a retrograde character [59]. Over the temperature range 593-900°C, it reaches its maximum
$(1.7 \cdot 10^{-2}$ at-% Li) at 800°C. The diffusion coefficients for lithium in germanium [2, 153] are
$D = 1 \cdot 10^{-9}$ cm^2/sec at 800°C and $D_0 = 2.5 \cdot 10^{-3}$ cm^2/sec, while the activation energy of diffusion
is Q = 49.4 kJ/mole (11.8 kcal/mole). The data obtained in Pell's investigation of solubility
[59] are in good agreement with the results of Morin and Reiss [155, 156], which are given in
Fig. 12. The heat of solution of lithium in solid germanium ΔH = 54.43 kJ/mole (13 kcal/mole),
while its temperature coefficient is $\Delta H/T$ = 19.68 kJ/mole·deg (4.7 kcal/mole·deg).

Sodium Germanides. The germanium-sodium system has been found to contain the
chemical compound NaGe (75.95% Ge), which was produced in the form of metallic needles by
synthesis at 650-700°C. A compound with the same composition was obtained by heating a mixture of powdered germanium and small chunks of sodium at 1000°C [3, 49]. The existence of
the germanium-rich compound $NaGe_4$ has been reported [48].

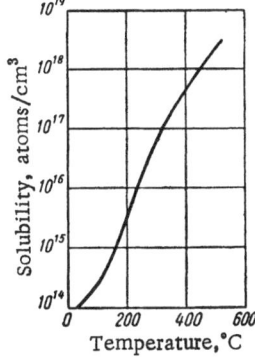

Fig. 12. Solubility of
lithium in germanium
at 20-500°C [155, 156,
170].

The NaGe obtained by the second method above was a hard,
dense substance with pyrophoric properties.

According to the data of Schaefer and Klemm [64], NaGe has
a monoclinic structure with the lattice constants a = 1.233, b = 0.67,
c = 1.142 nm, and β = 2.1 rad (120°).

The pycnometric density of this compound is 3.090 g/cm^3 and
its x-ray density is 3.140 g/cm^3.

The data of Witte and Schnering [157] on the crystal structure
of NaGe are in almost complete agreement with those of Schaefer
and Klemm [64]. They confirmed that the compound has a monoclinic lattice (C_{2h}^5−P21/c space group) with the constants a = 1.233,
b = 0.670, c = 1.142 nm, and β = 2.09 rad (119.9°).

Potassium Germanides. The K−Ge system has been
found to contain the chemical compounds KGe (64.99% Ge) and

KGe$_4$ (88.13% Ge). The compound KGe was produced by direct synthesis from potassium and germanium, while KGe$_4$ was obtained by thermal decomposition of KGe under high vacuum [49]. Johnson [3] prepared KGe in the form of crystals of indeterminate shape by heating powdered germanium and small chunks of potassium at 1000°C. Both compounds have a cubic structure with the lattic constant a = 1.278 nm for KGe and a = 1.59 mn for KGe$_4$ [64].

The pycnometric density of KGe is 2.780 g/cm^3 and its x-ray density is 2.850 g/cm^3.

Rubidium and Cesium Germanides. These systems have been found to contain the chemical compounds RbGe (45.93% Ge), RbGe$_4$ (77.25% Ge), CsGe (35.33% Ge), CsGe$_4$ (68.6% Ge). The compounds RbGe and CsGe were produced by direct synthesis at 600°C, while RbGe$_4$ and CsGe$_4$ were obtained by thermal decomposition of RbGe and CsGe respectively under high vacuum [49].

The compound RbGe and CsGe have a cubic structure with the lattice constants a = 1.319 and a = 1.367 nm respectively [64].

The x-ray densities of these compounds are 3.630 and 4.280 g/cm^3, while their x-ray densities are 3.660 and 4.280 g/cm^3.

The compound R̊bGe$_4$ also has a cubic structure with the constant a = 1.4 nm.

GERMANIDES OF COPPER-SUBGROUP METALS

Copper Germanides. This system was first studied by thermal and microscopic analysis [92]. Its phase diagram is given in Fig. 13 and shows the existence of α-, β-, γ-, δ-, ϵ-, η-, and ϑ-phases. The α- and ϑ-phases are solid solutions based on copper and germanium respectively. The β-phase is formed by a peritectic reaction at 820°C, while the γ-phase [a solid solution based on the compound Cu$_3$Ge (27.58% Ge)] is formed by a peritectic reaction at 744°C. The γ-phase is stable only at temperatures above 558°C and is converted to the low-temperature δ-phase modification at this temperature. At 700°C, the peritectic reaction $\gamma + 1 \rightleftharpoons \epsilon$ results in formation of the ϵ-phase, which is a eutectic with a ν-phase at 38% Ge and 650°C. The ϵ-phase undergoes a polymorphic transformation to the η-phase at 615°C; the latter is stable below this temperature. Since the microscopic study was conducted with unannealed, so-called nonequilibrium alloys, the concentration of the α-solid solution and the phase boundaries were only approximately established.

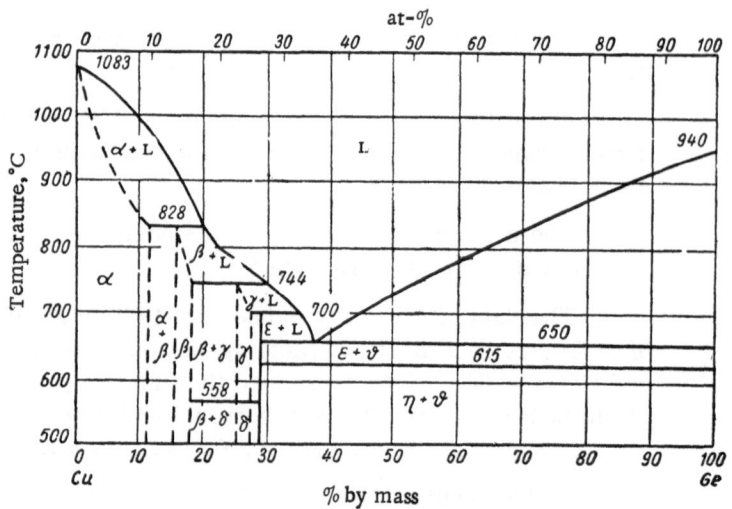

Fig. 13. Phase diagram of the copper – germanium system [92].

Weibke [93] showed that about 11.2% (10 at-%) Ge is dissolved in copper, the primary solid solution being in equilibrium with an intermediate ξ-phase (represented by the β-phase in Fig. 13), which contains 14.6-17.65% (13-16 at-%) Ge. When the germanium content is raised, the ξ-phase is converted to the γ-phase by a peritectic reaction; at 560°C, the latter phase undergoes eutectoid decay to the ξ- and ε-phases, but it is not subject to polymorphic transformation.

A group of researchers employed x-ray diffraction analysis to establish the boundary of the α-phase [94, 96]. The results of their solubility determinations are given in Table 3.

Table 3. Solubility of Germanium in Copper

Temperature, °C	Solubility of germanium in copper (%), from data in	
	[95]	[96]
822.5	13.44—13.56	—
821	—	12.9
300	11.0	—
200	—	9.68

Hume-Rothery [94] investigated the section of the diagram covering Ge contents between 21.2 and 22.25% and established that the ξ-phase has a broader homogeneity region than was previously supposed.

Schubert and Brandauer [97, 98] employed microscopic and x-ray analysis to check the system at Ge contents between 22.2 and 38.2% (20 and 35 at-%). X-ray diffraction analysis of alloys prepared by fusion of the electrolytic metals in evacuated quartz ampules was carried out at room and elevated temperatures. It was established the the ε-phase (γ) melts congruently at 755°C and gives a eutectic with the ξ-phase at 746°C. The ε-phase undergoes eutectoid decay to the ξ- and ε₁-phases at 570°C and forms the η-phase by a peritectic reaction at 700°C. The η-phase forms a eutectic with the germanium at 640°C and enters into a peritectic reaction with the ε-phase at 635°C, the latter resulting in formation of the ε₁-phase; the η-phase is stable only at high temperatures and is converted to the ε₁-phase below 612°C.

The data described above [93-98] have made it possible to refine the diagram proposed by Schwarz and Elstner [92]. Figure 14 shows such a refined phase diagram of the germanium—copper system. According to this diagram, α-, ξ-, ε-, ε₁-, η-, and Ge phases are present in

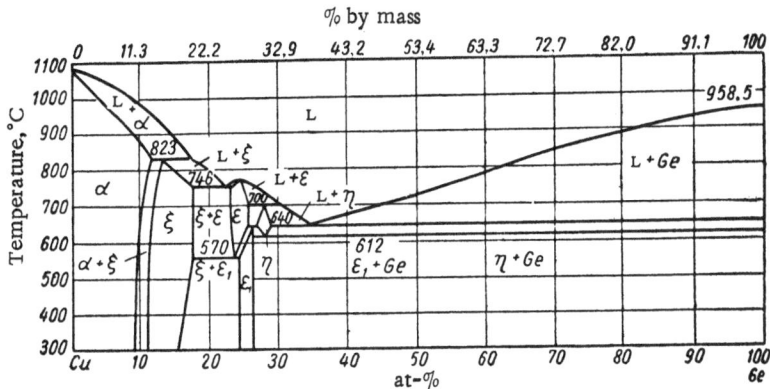

Fig. 14. Refined phase diagram of the copper−germanium system [170].

the system. The α-phase, which is a solid solution of germanium in copper, has a face-centered cubic lattice, whose constant increases with the germanium content, rising from 0.3616 for pure copper to 0.3629, 0.364, and 0.3651 nm for alloys containing 4.41, 7.78, and 10.81 at-% Ge respectively [158]. The ξ-solid solution is based on the compound Cu_5Ge (18.59% Ge) and has a hexagonal structure of the Mg type [93, 97, 98] having the lattice constants a = 0.2515, c = 0.411 nm, c/a = 1.638; a = 0.2614, c = 0.423 nm, and c/a = 1.622; and a = 0.266, c = 0.428 nm, and c/a = 1.612 at Ge contents of 11.9, 18.0, and 19.3 at-% (13.4, 20.06, and 21.4%) respectively.

According to Schubert and Brandauer [97, 98], the lattice constants of the ξ-phase at a Ge content of 14.3 at-% are a = 0.258, c = 0.421 nm, and c/a = 1.632. Hume-Rothery et al. [159] consider the ξ-phase to be an electronic compound with an electron : atom ratio of 3 : 2.

Two phases exist in the vicinity of the compound Cu_3Ge: the ε-phase is stable at high temperatures and the ε_1-phase is stable below 635°C. According to Schubert and Brandauer [97, 98], the ε-phase has a densely packed hexagonal structure. The lattice constants of this phase are a = 0.4168, c = 0.500 nm and c/a = 1.20 at a Ge content of 25.6% (23.2 at-%) and a temperature of 610°C, and a = 0.4168, c = 0.491 nm, and c/a = 1.18 at a Ge content of 27.6% (25.1 at-%) and a temperature of 700°C. The ε-phase forms a eutectic with the ξ-phase at a Ge content of 25.9% (23.5 at-%) and a temperature of 746°C. The ε_1-phase has a rhombically distorted hexagonal structure isomorphic with Ag_3Sn. The lattice constants of this phase at a Ge content of 25 at-% are a = 0.2645, b = 0.4549, and c = 0.419 nm. According to the data of Nowotny and Bachmayer [160], the ε_1-phase has a monoclinic structure with the lattice constants a = 0.2631, b = 0.420, c = 0.456 nm, and β = 1.56 rad (89° 41'). It can be regarded as a monoclinically deformed rhombic lattice. The η-phase has a cubic structure similar to that of β-W. A eutectic is formed between the η-phase and the germanium at a Ge content of 38.6% (36.0 at-%) and a temperature of 640°C.

The solubility of copper in solid germanium is very low. As was established by electrical-conductivity measurements and by the radioactive-isotope method, the maximum solubility of copper in germanium at 600°C is about $5 \cdot 10^{14}$ atoms/cm³ [162].

The heat of solution of copper in solid germanium is ΔH = 171.66 kJ/mole (41 kcal/mole), while the temperature coefficient is $\Delta H/T$ = 62.80 kJ/mole·deg (15 kcal/mole·deg). The solubility of copper in germanium increases as the temperature is raised to 880°C, reaching $3.6 \cdot 10^{16}$ atoms/cm³. A further rise in temperature and approximation to the melting point of the germanium leads to a decrease in maximum solubility. According to the data of other researchers [161, 164], the solubility amounts to $7 \cdot 10^5$ at-%; the maximum is one copper atom for each 10^7 germanium atoms [165].

The literature contains no reliable data on the diffusion rate of copper in germanium at temperatures below 700°C. Diffusion of copper in solid germanium is very rapid at temperatures above 750°C and does not obey Ficke's law. The high diffusion rate of copper in germanium can be explained by assuming that the copper atoms are displaced along the interstices of the germanium crystal lattice. Lying in the interstices of the lattice, these atoms can recombine with the vacancies and occupy lattice points. The vacancies that participate in this process are usually located near the surface of the germanium crystal or near dislocations. The substantial influence of the structural ideality of a germanium crystal on the rate at which copper diffuses in it can also be explained in this manner.

The data of Reynolds and Hume-Rothery [166] are in good agreement with the phase diagram in Fig. 14. These authors prepared alloys from high-purity metals (the germanium contained less than 0.0005% impurities and the copper was 99.998% pure) by fusion in graphite crucibles under a potassium chloride flux. The principal difference between their phase diagram [166] and that in Fig. 13 is that the boundaries of the ε_1-, ε-, and η-phases lie at higher germanium contents (1-2 at-%). Reynolds and Hume-Rothery believe that this discrepancy resulted

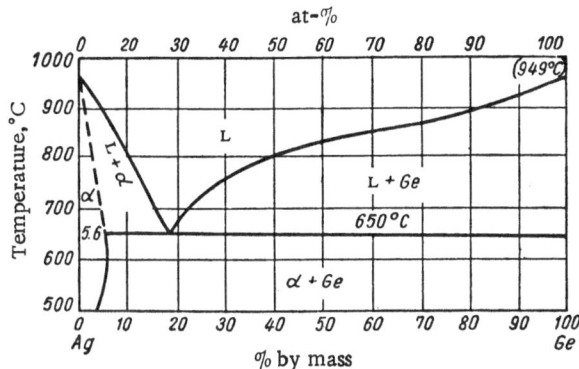

Fig. 15. Phase diagram of silver–germanium
system [168].

from loss of germanium during annealing of the powders before the x-ray diffraction analysis
was made.

Addition of germanium improves the mechanical properties of copper and increases its
hardness and strength. Alloys become more brittle. The modulus of elasticity of copper–
germanium alloys decreases from about 131.000 to about 119.000 MN/m^2 (from 13.100 to 11.900
kg/mm^2) as the germanium content is raised from 0 to 9.14% [167]. Annealed bars 6 mm in diameter, produced by rolling and cold drawing, were tested. The resistivity of copper increases
from 1.55 μohm·cm for pure copper to 9.2 μohm·cm for an alloy containing 2.39% (2.1 at-%) Ge.

Silver Germanides. The phase diagram of the germanium–silver system (Fig. 15)
was first constructed by Briggs et al. [168] on the basis of thermal and microscopic analysis
and was refined by Hume-Rothery et al. [94]. Germanium–silver alloys contain a eutectic that
melts at 651 ± 0.5°C. The eutectic Ge concentration is 19% (26 at-%). The data on the solubility
of germanium in silver at temperatures below 600°C are contradictory. Thus, Hume-Rothery
established that alloys containing less than 15% (20 at-%) Ge had a monophasic α-solid solution
structure after annealing at 500, 404, and 298°C for 15, 24, and 46 days respectively.

The solubility (in %) of germanium in silver at different temperatures has been established by
x-ray analysis [96, 170]. The data obtained are given below:

Temperature, °C	575	470	375	308	270
Solubility					
[96].................	7.9	6.7	3.8	2.7	1.5
[170]................	5.46	3.9	2.72	1.96	1.03

The results of Nowotny and Bachmayer [160], who showed the solubility of germanium in
solid silver at 650°C to be 5.6%, are in good agreement with these data. The determinations
were made by microscopic analysis. The solubility of silver in solid germanium is low. Bugai
et al. [169] found the maximum solubility at 675°C to be 1·10^{15} atoms/cm^3. The solubility of silver in germanium has a retrograde character (Fig. 16). The germanium, entering into a solid
solution, increases the lattice constant of the silver. The change in the lattice constant of the
solid solution in Ag–Ge alloys is shown below [169]:

Ge content, at-%..........	0	1.56	3.02	4.61	7.16
a, nm.................	0.4085	0.4086	0.4086	0.4086	0.4087

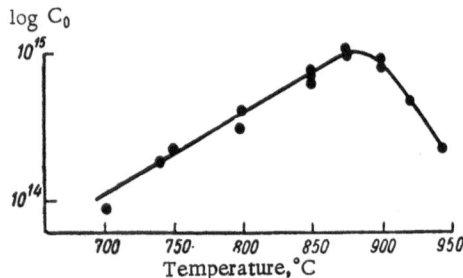

Fig. 16. Solubility of silver in germanium [169].

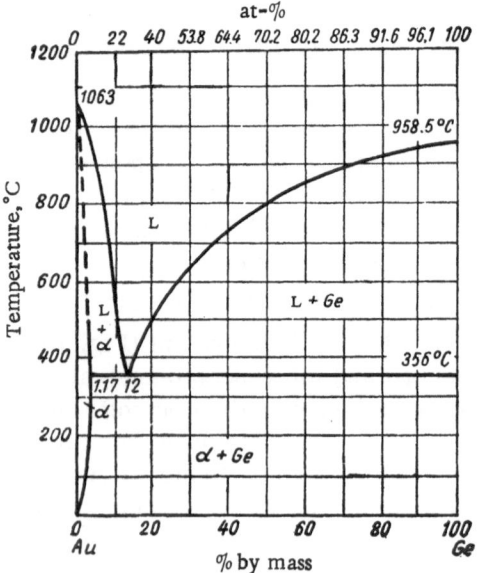

Fig. 17. Phase diagram of the gold–germanium system [172, 173].

Silver in germanium has three acceptor levels, which are located at: a) $0.20 \cdot 10^{-19}$ J (0.13eV) above the upper boundary of the valence zone; b) $0.44 \cdot 10^{-19}$ J (0.28 eV) below the conduction zone; c) $0.144 \cdot 10^{-19}$ J (0.09 eV) below the conduction zone.

These acceptor levels are intermediate in position between those introduced into the energy spectrum of germanium by other triple acceptors (copper and gold). The introduction of three acceptor levels into germanium by silver is readily explained by the tetrahedral-bond model, since the abbreviated formula for the electronic structure of silver has the form $4d^{10}5s^1$. The silver atoms are apparently located at the points of the germanium lattice, forming a solid substitution solution. The diffusion constant of Ag in Ge is D = $9 \cdot 10^7$ cm^2/sec at 800°C [169] and D = $5 \cdot 10^{-7}$ cm^2/sec at 710°C [171]. Alloys of germanium with silver have low electrical resistance. The heat of solution silver in solid germanium is ΔH = 234.46 kJ/mole (56 kcal/mole), while its temperature coefficient is $\Delta H/T$ = 83.74 J/mole·deg (20 cal/mole·deg) [59].

Gold Germanides. The phase diagram of the germanium—gold system was investigated by Jaffee et al. [172, 173], who employed thermal, microscopic, and x-ray analysis and hardness measurements. This diagram is of the simple eutectic type (Fig. 17). The eutectic contains 73 at-% gold at 356°C.

The solubility of germanium is solid gold is comparatively high and reaches 3.2 at-% (1.2%) Ge at the eutectic temperature. According to other data [175], it varies with temperature. Data on the solubility of germanium is gold as a function of temperature are given below:

Temperature, °C	450	400	363	350	300	250	200
Maximum solubility of Ge in Au:							
at-%	2.8	3.0	3.12	2.2	0.9	0.4	0.2
by mass	1.05	1.13	1.18	0.83	0.33	0.15	0.07

The maximum solubility of gold in germanium is about $1 \cdot 10^{-5}$–$2 \cdot 10^{-6}$ at-% [172, 173, 450]. The diffusion constant of gold in germanium and the activation energy of diffusion are D_0 = 12.6 cm^2/sec and Q = 217.71 kJ/mole (52 kcal/mole) [2]. Addition of germanium increases the hardness and strength of gold, but it also makes alloys more brittle. Alloys of gold with germanium containing up to 5% Ge have adequate malleability. Any further increase in germanium content makes the alloys brittle. The eutectic alloy has a Vickers hardness of about 2000 MN/m^2 (200 kg/mm^2) and an ultimate strength of about 391 MN/m^2 (39.1 kg/mm^2). Its relative elongation is extremely low. Addition of 0.25 and 0.53% Ge causes the resistivity of gold to change from 2.205 to 5.792 and 9.174 μohm·cm respectively. Crystallization of eutectic alloys with a higher germanium content entails an increase in volume. The eutectic structure is extremely fine-grained in preeutectic alloys of eutectic composition and coarse-grained in eutectic alloys. Gold-rich alloys retain the color of this metal until a Ge content of 12% is reached [172, 173].

Fomin et al. [188] recently studied the distribution of gold as an alloying additive in germanium monocrystals. They established that, because of the low diffusion rate of gold in germanium and the high crystallization rate of the melt, the gold is nonuniformly distributed over the length and cross section of the ingot, collecting in eutectic inclusions along definite crystallographic planes. According to x-ray diffraction data, these planes are the {111} planes.

GERMANIDES OF ALKALINE-EARTH METALS

Beryllium Germanides [237]. There is almost no information on this system in the literature. A beryllium−germanium alloy containing 10% Ge and cooled slowly from the molten state has a diphasic structure. Metallographic investigation of the structure of this alloy showed that there were light-colored inclusions of a chemical compound along the grain boundaries and within the grains [177].

Belyaev and Zhidkov [178] studied the diffusion of beryllium in monocrystalline germanium alloyed with antimony. For this purpose, a layer of beryllium up to 10 μ thick was vacuum-deposited on the surface of germanium specimens in the form of parallelepipeds $2 \times 3 \times 10$ mm in size. The specimens were then annealed in a quartz tube at a pressure of 133.3 mN/m² (10⁻³ mm Hg) and a temperature of 920°C for 24-150 h. The depth to which the beryllium diffused in the germanium was determined from the position of the p-n junctions and amounted to 15-70 μ.

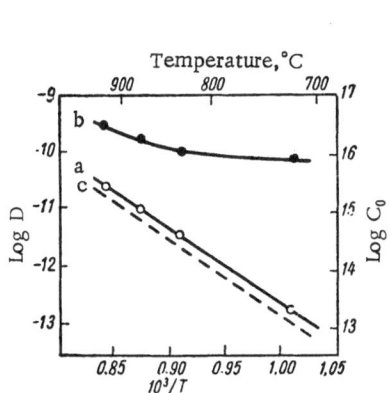

Fig. 18. Diffusion constant and maximum solubility of beryllium in germanium as functions of temperature [178].

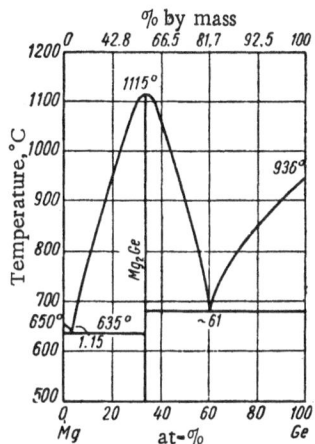

Fig. 19. Phase diagram of the magnesium−germanium system [99].

Figure 18 shows the diffusion constant (a) and maximum solubility of beryllium (b) and zinc (c) in germanium as functions of temperature. The aforementioned authors established that the solubility of beryllium in germanium depends little on temperature over the range investigated, while the diffusion constant is described by the equation

$$D = 0.5 \exp\left(-\frac{2.5}{kT}\right).$$

Magnesium Germanides. The phase diagram of the germanium—magnesium system was worked out by Klemm and Westlinning [99], who employed thermal and microscopic analysis. As can be seen from this diagram (Fig.19), the chemical compound Mg_2Ge (59.88% Ge) is formed in this system, melting congruently at 1115°C and forming a eutectic with the magnesium at a Ge content of 3.55% (1.2 at-%) and a temperature of 635°C and with the germanium at a Ge content of 82.2% (60.8 at-%) and 690°C. Measurement of the lattice constants showed that the solubility of germanium in magnesium is slight at 600°C (0.097%) and negligible at lower temperatures. No change in the lattice constant of magnesium in its alloys with germanium was detected at 450°C. Thermal analysis showed the $Mg-Mg_2Ge$ eutectic to have the coordinates 1.15 at-% (5.4%) Ge and 634.7°C [179].

Inclusions of the compound Mg_2Ge are observed microscopically along the grain boundaries of alloys quenched from 600°C at a Ge content of 0.3% [99]. The existence of the compound Mg_2Ge in this system was confirmed by Savitskii and Baron [180].

The compound Mg_2Ge has a cubic structure of the CaF_2 type with the lattice constant a = 0.6386 nm [100] (0.6380 nm [99]).

The compound Mg_2Ge was produced by fusing the components at 1200-1300°C; an alloy of nonhomogeneous composition was obtained at lower temperatures [179].

The density of Mg_2Ge is 3.086 g/cm^3 [100]. This compound is a semiconductor with a forbidden-zone width of $1.18 \cdot 10^{-19}$ J (0.74 eV) [184]. According to Mooser and Pearson [492], Mg_2Ge satisfies all the conditions for formation of semiconductive phases, as has been confirmed by investigation of its electrophysical properties [184]. In accordance with the electron-valence bonding pattern for Mg_2B^{IV} compounds, each germanium atom forms four covalent sp^3 bonds, which can undergo rotary resonance among eight positions as a result of the presence of unfilled orbitals in the magnesium atoms. Each germanium atom thus forms eight half-bonds with the magnesium atoms surrounding it. The s and p subshells of the magnesium atoms are only half-filled. At the same time, each magnesium atom utilizes two of its electrons to saturate the bonds with four adjacent germanium atoms, so that no electrons remain for formation of $Mg-Mg$ bonds.

Glazov and Glagoleva [493] believe that homeopolar bonding is the principal type of chemical bonding in solid Mg_2Ge, which serves as indirect confirmation of the validity of the bonding pattern described above for this compound.

Investigation of the temperature function of electrical conductivity for solid and liquid Mg_2Ge established that its conductivity increases abruptly on melting, to about 10^4 ohm-1 · cm-1; the character of its chemical bonding changes from predominantly covalent to metallic in this case.

Raynor [179] determined the lattice constants of an alloy containing 0.1% Ge and found them to be a = 0.320 nm and c/a = 1.623 after annealing at 602°C for 18 days and a = 0.3209 nm and c/a = 1.623 after annealing at 450°C for 10 days. The constants were determined at 20°C. Addition of 0.35% germanium markedly increases the strength of magnesium. When the germanium content is raised from 44 to 68%, the hardness increases from about 220 to 680 MN/m^2 (from 22 to 68 kg/mm^2) (the hardness was measured by indentation with a pobedite taper). The hardness of an alloy containing 72% Ge is about 730 MN/m^2 (73 kg/mm^2) (these curves are taken from graphs given by Brauer and Tisler [100]). The microhardness of the eutectic has been found to be about 660 MN/m^2 (66 kg/mm^2) and that of the chemical compound Mg_2Ge to be about 3300-3500 MN/m^2 (330-350 kg/mm^2). The germanium—magnesium alloys studied by Savitskii and Baron [180] were brittle at room temperature and softened on heating, their strength undergoing a slight initial increase.

Eremenko and Lukashenko [185] measured the emf of galvanic concentration cells to study the thermodynamic properties of Mg_2Ge at 427-627°C. The electrolyte was a eutectic KCl—

LiCl—MgCl$_2$ mixture (see [186, 187] for a description of the method). The alloys were prepared from magnesium (99.92% pure) and monocrystalline germanium. They were subjected to preliminary annealing at 627°C for 24 h. Since magnesium is insoluble in solid germanium and Mg$_2$Ge has no homogeneity region, the change in the isobaric-isothermal potential as a result of formation of one mole of Mg$_2$Ge from the pure components by the reaction $2Mg_S + Ge_S \rightleftharpoons Mg_2Ge_S$ was determined from the expression

$$\Delta Z^\circ_{Mg_2Ge} = -2nFE,$$

where n is the valence of the magnesium ions (+ 2), F is the Faraday constant, and E is the emf of the cell.

Figure 20 shows the results of the emf measurements for alloys containing 61 and 79 at-% Ge. The results were processed by the method of least squares, which yielded the following equation for the emf (V) as a function of temperature:

$$E = 0.272 - 0.38 \cdot 10^{-4}\ T$$

with an average deviation of no more than ± 0.0035 V.

The temperature function of the isobaric-isothermal potential of magnesium germanide formation over the temperature range investigated is described by the equation (kJ/mole)

$$\Delta Z^\circ_{Mg_2Ge} = -105.0 + 0.01464\ T.$$

The heat and entropy of formation are $\Delta H^\circ_{Mg_2Ge} = -35.0$ kJ/g-atom (-8.37 kcal/g-atom) and $\Delta S^\circ_{Mg_2Ge} = -4.88$ J/deg·g-atom (-1.17 cal/deg·g-atom).

According to more recent data [490], the heat of formation, free energy of formation, standard entropy of formation, and standard entropy of Mg$_2$Ge are $\Delta H^\circ = -9.18 \pm 0.02$ kcal/g-atom (-38.46 kJ/g-atom), $-\Delta F^\circ_{298} = 8.9 \pm 0.3$ kcal/g-atom (37.29 kJ/g-atom), $-\Delta S^\circ_{298} = 0.8 \pm 0.1$ eu/g-atom, and $S'^\circ_{298} = 6.9 \pm 1.0$ eu/g-atom.

Clung et al. [491] measured the rate of longitudinal and transverse sound propagation in the [100] and [111] directions and the thermal expansion of Mg$_2$Ge crystals over the temperature range 80-300°K. He established that the speed of sound depends very little on temperature or crystal orientation. The elastic constants calculated from these measurements had the values (in 10^{11} dyn/cm^2): $C_{11} = 11.79 \pm 0.15$, $C_{12} = 2.30 \pm 0.50$, and $C_{44} = 4.65 \pm 0.10$. The calculated Debye temperature was 492 ± 8°K, which is in good agreement with the experimental value of 488°K.

According to Labotz et al. [432], the electrophysical properties of Mg$_2$Ge include a forbidden-zone width of 0.70 eV, a coefficient of thermal conductivity of 0.0628 W/cm·deg, a resistance of 2-4 ohm·cm, and a thermal emf of +710 V/deg.

Calcium Germanides. The germanium-calcium system has been found to contain the chemical compounds CaGe$_2$ (70.36% Ge), CaGe (64.43% Ge), and Ca$_2$Ge (47.53% Ge) [55, 82, 104]. In addition to the aforementioned germanides, Helleis et al. [103] discovered the two additional compounds Ca$_{33}$Ge and Ca$_7$Ge, which had the form of cubic and acicular crystals. The cubic crystals contained 79.4% Ca, 20.6% Ge, 0.4% Fe and 0.1% C. They were assigned the formula Ca$_7$Ge (O^7_h-Fd3m space group).

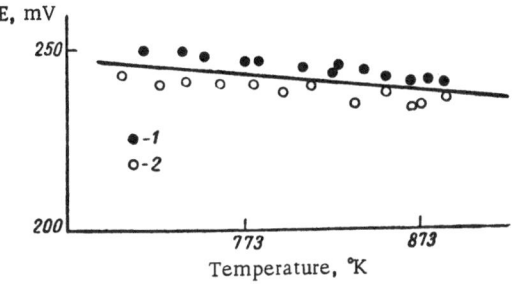

Fig. 20. Curve representing emf as a function of temperature [185] for alloys containing: (1) at-% and (2) 79 at-% Ge.

The acicular crystals contained 95% Ca, 5.0% Ge, 0.4% Fe, and 0.6% C. These crystals were assigned the formula $Ca_{33}Ge$ (O_h^5 - Fm3m space group). Table 1 presents data on the structures of these germanides.

According to Semenkovich [34], Ca_2Ge is a semiconductor with a forbidden-zone width of $2.72 \cdot 10^{-19}$ J (1.7 eV).

Strontium Germanides. According to the data of Merlo and Formasini [478], the compound S_2Ge has been found in the germanium−strontium system; it has a rhombic lattice of the CrB type (D_{17}^2-Cmcm space group) with the constants a = 0.482, b = 1.139, and c = 0.4167 nm.

Barium Germanides. A recently published article [438] reported the existence of the monogermanide BaGe. This compound was obtained by fusing barium (98.9% Ba) with germanium (99.9% Ge) in a purified-argon atmosphere at a pressure of 98.06 kN/m^2 (1 atm) in an arc furnace. X-ray diffraction analysis showed that the resultant compound belongs to the rhombic system of the CrB type with the lattice constants a = 0.507, b = 1.198, and c = 0.430 nm.

The fusibility diagram of the germanium−barium system (Fig. 21) was plotted by Andrianov et al. [71], with the aid of thermal analysis. These authors established that the system contains one congruently melting compound, BaGe (t_m = 1145°C), and two incongruently melting compounds, $BaGe_2$ (t_m = 1050°C) and Ba_2Ge (t_m = 940°C). According to x-ray diffraction data, $BaGe_2$ has a primitive cubic lattice (O^7-$P4_132$ space group) with the constant a = 1.452 nm and 28 formal units in the elementary cell. The experimental density of the compound is 4.28 g/cm^3.

Compounds of barium with germanium, like other compounds of alkali-earth metals with germanium, are characterized by an acrid odor, which is apparently due to liberation of germanium.

All compounds of barium with germanium have high hardness and readily form clusters of large crystals.

GERMANIDES OF ZINC-SUBGROUP METALS

Zinc Germanides. The phase diagram of the germanium−zinc system (Fig. 22) is of the simple eutectic type and was first worked out by Gebhardt [65], who employed thermal and microscopic analysis. The eutectic lies at 6% Ge and 398°C. The maximum germanium content in solid solution in the zinc at 380°C is less than 0.1%, while zinc has not been found to be soluble in solid germanium [65, 67].

A rise in the germanium content of zinc-rich alloys increases their hardness. Figure 23 shows the Brinell hardness of alloys cast in chill molds as a function of germanium content.

Boronin and Evseev [189] measured the pressure of saturated zinc vapor in order to study the thermodynamic properties of molten and solid Ge−Zn alloys over the temperature range 342-466°C. They established that germanium undergoes severe structural distortion in both the solid and liquid states when it forms solutions with zinc. The entire system exhibits great deviations from an ideal solution.

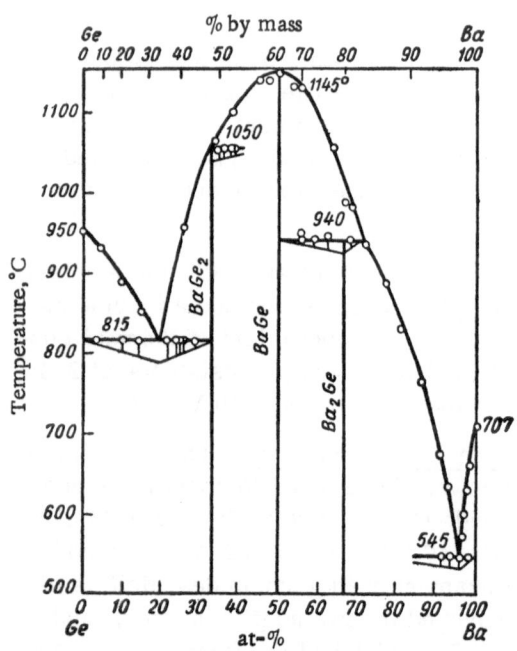

Fig. 21. Fusibility diagram of the barium−germanium system [71].

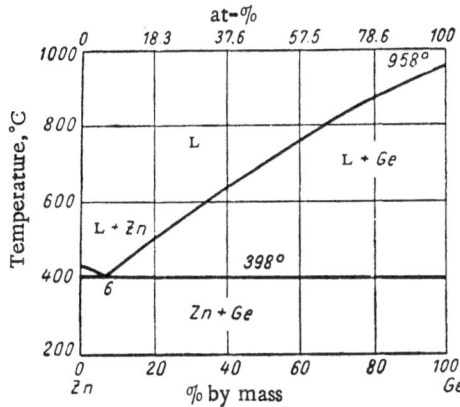

Fig. 22. Phase diagram of the zinc—germanium system [65].

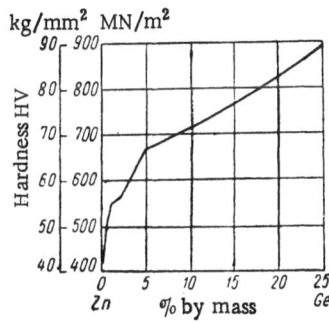

Fig. 23. Change in hardness of cast Zn—Ge alloys as a function of germanium content [65, 170].

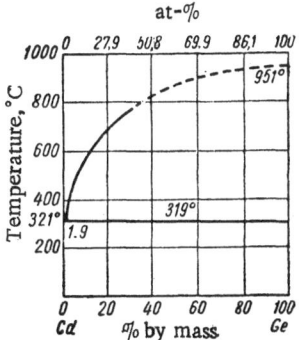

Fig. 24. Phase diagram of the cadmium—germanium system [66].

Cadmium Germanides. The phase diagram of the germanium—cadmium system was worked out by Spengler [66], who employed thermal and microscopic analyses. This phase diagram is shown in Fig. 24. Germanium and cadmium form a system of the simple eutectic type, with the eutectic lying at 1.9% Ge and 319°C. The solubility of cadmium in germanium is less than 0.1%. The alloys were prepared from 99.9% pure metals by fusion in porcelain crucibles in an inert atmosphere.

The liquidus for alloys containing more than 35% Cd could not be determined, because of the considerable volatility of cadmium at these temperatures. The solubility of cadmium in germanium at 840°C is $2 \cdot 10^{18}$ atoms/cm^3 [174].

Mercury Germanides. The soubility of germanium in liquid mercury is very low and becomes noticeable only above 250°C. Electrical-resistance measurements have shown that at least 0.027% Ge dissolves in mercury at 300°C [190, 312].

CHAPTER 4

TRANSITION-METAL GERMANIDES

GERMANIDES OF RARE-EARTH METALS

Little research has been done on compounds of rare-earth metals with germanium. Digermanides with a tetragonal structure of the α-ThSi$_2$ type are known for the elements of the cerium subgroup, including LaGe$_2$ [195], CeGe$_2$ [191], PrGe$_2$ [198], NbGe$_2$ [192, 195], and SmGe$_2$ [195]. The digermanide EuGe$_2$ has a crystal structure of a different type ($\bar{P3}$mI-D3$_{3d}^3$ space group), the AlB$_2$ type [193].

The digermanides LaGe$_2$, NdGe$_2$, SmGe$_2$, and GdGe$_2$ are isostructural with CeGe$_2$ and PrGe$_2$ (the α-ThSi$_2$ structural type, 14/amd-D$_{4h}^{19}$ space group) and their lattice constants are given in Table 1 [146].

The x-ray diffraction patterns of cast and annealed alloys of lanthanum and cerium with germanium having MeGe$_2$ compositions exhibit splitting of certain lines, which indicates rhombic deformation of the structure.

By analogy with the silicides of rare-earth metals of the cerium subgroup, it is assumed that the compounds LaGe$_2$ and CeGe$_2$ have low-temperature modifications with a structure of the α-CdSi$_2$ type (Imma-D$_{2h}^8$ space group). The lattice constants of LaGe$_2$ and CeGe$_2$ are a = 0.441, b = 0.430, and c = 1.419 nm and a = 0.436, and b = 0.426, and c = 1.407 nm respectively [145, 146].

Alloys with the composition RGe$_2$ are nonhomogeneous for the metals of the yttrium subgroup (Tb, Dy, Ho, Er, Tu, and Lu). The lattice constants and relative reflection intensities enable us to assume a crystal structure of the AlB$_2$ type (P6/mmm-D$_{6h}^1$ space group) for compounds of these metals. Phase analysis of additive alloys containing 60 and 50 at-% Ge showed that similar compounds are formed in alloys of all the aforementioned yttrium-subgroup metals with germanium and that their compositions are close to R$_2$Ge$_3$. Their lattice constants are given in Table 1.

All the structural types of rare-earth digermanides are interrelated. The coordination number of the germanium atoms is nine in all four structures. The nearest neighbors of each germanim atom are three other germanium atoms at distances close to the d$_{Ge-Ge}$ in the Ge structure (0.244 nm) or at shorter distances (e.g., in Er$_2$Ge$_3$), which indicates that there is covalent bonding between the germanium atoms.

The transition from the AlB$_2$ to the EuGe$_2$ type is associated with a shift in the germanium atoms in the germanium coordination sphere along the L$_3$ axis and a simultaneous increase in the distance between the rare-earth atoms lying on this axis (a rise in c/a). The transition from the AlB$_2$ type to the α-ThSi$_2$ type is associated with deformation of the trigonal prisms; the base of the prism becomes an isosceles triangle, while one rectangular lateral face becomes a square. In the α-GdSi$_2$ structure, all the lateral faces of the prism are rectangular [146].

The germanium–praseodymium system has been the most completely studied and has been found to contain the compound PrGe (34.0% Ge) in addition to the aforementioned digermanide (50.75% Ge). This compound has a rhombic structure of the CrB type with the lattice constants a = 0.4474, b = 1.1098, and c = 0.4064 nm [196]. The elementary cell contains four molecules of the compound. Its density is 6.780 g/cm^3.

According to Iandelli [196], $PrGe_2$ has a tetragonal structure of the $ThSi_2$ type with the lattice constants a = 0.4253, c = 1.3940 nm, and c/a = 3.277. The elementary cell also contains four molecules of the compound. Its density is 7.24 g/cm^3.

According to Gladyshevskii et al. [192], the compound $NdGe_2$ is isostructural with α-$ThSi_2$ and has the lattice constants a = 0.4258, c = 1.387 nm, and c/a = 3.26.

According to Gladyshevskii [193], the compound $EuGe_2$ has a hexagonal structure with the lattice constants a = 0.4102, c = 0.4995 nm, and c/a = 1.218. The pycnometric density of this compound is 6.67 g/cm^3.

The structure of scandium and yttrium digermanides was determined by Schob and Parthe [198], who established that $ScGe_2$ has a lattice of the $ZrSi_2$ type with the constants a = 0.3888, b = 1.4873, and c = 0.3793 nm. The compounds YGe_2 and Y_2Ge_3 in the yttrium−germanium system have lattices of the $ThSi_2$ and AlB_2 types with the constants a = 0.4060, c = 1.368 nm, and c/a = 3.397 and a = 0.3935, c = 0.4139 nm, and c/a = 1.052 respectively.

Arbuckle and Parthe [143] employed the x-ray diffraction method to study the compounds La_5Ge_3, Ce_5Ge_3, and Sc_5Ge_3. These compounds have a hexagonal structure of the Mn_5Si_3 type ($D8_8$) with the lattice constants a = 0.8958, c = 0.6795 nm, and c/a = 0.759; a = 0.8875, c = 0.657 nm, and c/a = 0.740; and a = 0.7939, c = 0.5883 nm, and c/a = 0.741 respectively. The x-ray density of La_5Ge_3 is 3.72 g/cm^3, that of Sc_5Ge_3 is 2.76 g/cm^3, and that of Ce_5Ge_3 is 3.92 g/cm^3.

The compounds La_4Ge_3 and Pr_4Ge_3 have recently been obtained [440]. They have a cubic structure of the Th_3P_4 type with the constants a = 0.9356 and 0.9153 nm respectively.

Lyutaya and Goncharuk [147] investigated the Ge−La system and, in addition to the compounds La_5Ge_3 and $LaGe_2$, established the existence of lanthanum monogermanide, LaGe, which has a structure of the FeB type and the constants a = 0.845, b = 0.413, and c = 0.610 nm.

Various researchers have studied the structures of rare-earth monogermanides [199, 200, 203]. They have shown that these structures exhibit rhombic syngony of the FeB and CrB types. However, according to the data of Hohnke and Parthe [434] and of Tharp et al. [435], the monogermanides of gadolinium and lutecium must be assigned a body-centered tetragonal structure, whose lattice constants are given in Table 1.

Using x-ray diffraction analysis, Baenziger and Hegenbarth [142] found the compounds Cd_5Ge_3 and Dy_5Ge_3, which have a hexagonal lattice of the Mn_5Si_3 type with the constants a = 0.8546, c = 0.641 mm, and c/a = 0.750 and a = 0.8438, c = 0.6336 mm, and c/a = 0.750 respectively.

Parthe [141] investigated the germanium−yttrium system. Samples were prepared by arc-fusion in a helium atmosphere and subjected to x-ray diffraction analysis by the powder method. The existence of the compound Y_5Ge_3 (37.4 at-% Ge) was established from the x-ray data; it has a hexagonal lattice of the $D8_8$ type and the constants a = 0.8471, c = 0.635 nm, and c/a = 0.749 (D_{6h}^3-$P6_3$/mcm space group). The yttrium atoms are in the 6(d) and 6(g)$_I$ positions with x_1 = 0.25, while the germanium atoms are in the 6(d)$_{II}$ positions with x = 0.61.

The compound Y_5Ge_3, like Y_5Si_3, has an unusually high c/a ratio, which is due to electron transfer from the atoms in the 6(g)$_I$ position to those in the 4(d) position and a reduction in the distance between the atoms in these positions.

The x-ray density of Y_5Ge_3 is 5.57 g/cm^3.

Gladyshevskii and Burnashiva [201], who studied the gadolinium−germanium system, confirmed the existence of the compounds Gd_5Ge_3, GdGe, $GdGe_3$, and $GdGe_2$, and established that of the new compound $GdGe_{2-n}$. They investigated alloys containing 25–85 at-% Ge, which were prepared by arc-fusion of high-purity metals in an inert atmosphere. The specimens were vacuum-annealed at 600°C for 350 h and then subjected to x-ray and microstructural phase analysis.

The compound richest in gadolinium has the composition Gd_5Ge_3. This compound is iso-structural with the previously described germanides La_5Ge_3 and Ce_5Ge_3 [143], as well as with Y_5Ge_3 [141], and has a hexagonal lattice of the Mn_5Si_3 type (D_{6h}^3 space group) with the constants a = 0.858, c = 0.645 nm, and c/a = 0.751, which is in rather good agreement with the results obtained by Baenziger and Hegenbarth [142].

The compound GdGe is of the CrB structural type (D_{2h}^7-Cmcm space group). However, the lattice constants of this compound (a = 0.433, b = 1.079, and c = 0.398 nm) differ considerably from those given by Moriarty and Baenziger [204] (a = 0.4175, b = 0.3960, and c = 1.061 nm) and are closer to the results obtained by other authors [199, 200, 203]. The differences in the lattice constants are outside the limits of experimental error and are due to the different degrees of purity of the initial compounds employed in the various studies.

The compound Gd_2Ge_3 has a hexagonal structure of the defective AlB_2 type (D_{5h}-P6/mmm space group) with the lattice constants a = 0.3973, c = 0.4179 nm, and c/a = 1.052. It is related to the previously discovered R_2Ge_3 compounds (where R is Tb, Dy, Ho, Er, Tu, Yb, or Lu) with defective AlB_2 structures [146].

Cast alloys containing more than 66.7 at-% Ge exhibit the compound $GdGe_3$, whose crystal structure has not as yet been established. This compound decomposes after prolonged (350 h) annealing at 600°C. The digermanide $GdGe_2$, with an α-$ThSi_2$ structure, has not been found in annealed alloys. The most germanium-rich compound at 600°C has the composition $GdGe_{2-n}$ and a crystal structure of the α-$GDSi_2$ type (D_{2h}^{28}- Imma space group) with the constants a = 0.4130, b = 0.4096, and c = 1.376 nm. The existence of a compound with an α-$GdSi_2$ structure in cast alloys containing a small excess of gadolinium indicates that the hexagonal structure is converted to a rhombic structure as a result of changes in temperature or in compound composition (the transition from $GdGe_2$ to $GdGe_{2-n}$, which is accompanied by removal of germanium atoms). The Gd—Ge system is thus analogous to the Gd—Si system. All gadolinium germanides (with the exception of GdGe) are isostructural with the corresponding silicides.

Moriarty et al. [205] and Smith et al. [206] investigated the crystal structure of ytterbium (Yb_3Ge_5), holmium (Ho_5Ge_3), and erbium (Er_5Ge_3) germanides. The compound Yb_3Ge_5 is iso-structural with Th_3Pt_5 and Th_3Pd_5 and has a hexagonal lattice (P62m space group) with the constants a = 0.603 and c = 4166 nm; its x-ray density is 8.77 g/cm^3 [206]. The compounds Ho_5Ge_3 and Er_5Ge_3 have a hexagonal lattice of the Mn_5Si_3 type (D_{6h}^3 space group) with the constants a = 0.8410 and c = 0.630 nm and a = 0.8367 and c = 0.6266 nm respectively [205].

Gladyshevskii [202] suggested that, in addition to the aforementioned germanides with the composition R_5Ge_3, other alloys of rare-earth metals with germanium also contain compounds with the same composition. In order to investigated this problem, rare-earth metals (La, Ce, Pr, Nd, Sm, Eu, Gd, Tb, Dy, Ho, Er, Tu, Yb, and Lu) with a base-metal content of 97–99% were fused in a helium atmosphere in an arc furnace. The resultant alloys were annealed at 500°C for 50 h and subjected to x-ray diffraction analysis by the powder method. The investigation showed that the alloys, except for those of Eu and Tb, were R_5Ge_3 compounds with an Mn_5Si_3 structure. Table 1 gives their lattice constants [202]. Europium and ytterbium do not form compounds with an Mn_5Si_3 structure. Compounds with an undetermined structure were detected in alloys with the composition investigated or compositions close to it.

However, x-ray diffraction analysis [441] of monocrystals with the approximate composition Sm_5Ge_3 showed that the compound Sm_5Ge_4, having a rhombic structure, was present instead of Sm_5Ge_3. Compounds with this composition have also been found in alloys of germanium with neodymium, gadolinium, terbium, erbium, and yttrium. Table 1 gives the lattice constants of these compounds.

Paderno et al. [76] investigated a number of the physical properties of lanthanum digermanide with specimens obtained by sintering the components and then fusing them in an argon atmosphere. Its resistance at room temperature was 659 μohm·cm, its microhardness was 3750 MN/m^2 (375 kg/mm^2), and its coefficient of thermal expansion over the temperature range 0-800°C was 8.9·10^{-6} deg^{-1}. According to the data of Matthias et al. [207], the compounds $NdGe_2$ and $PrGe_2$ have ferromagnetic properties. There have also been more recent studies of rare-earth germanides [451-454].

GERMANIDES OF SUBGROUP-IVa METALS

<u>Titanium Germanides</u>. X-ray diffraction analysis has established the existence of the compounds Ti_5Ge_3 (47.63% Ge), $TiGe_2$ (75.19% Ge)[82, 105, 106], and TiGe [211] in this system.

McQuillan [208] employed microscopic analysis at temperatures between 500 and 1500°C to study the structure of alloys containing 0-16% Ge. The alloys were prepared from titanium iodide and 99.9% germanium by fusion in a water-cooled copper crucible in an argon atmosphere in an arc furnace. Heat treatment was carried out in a vacuum. The holding times at different temperatures are given below.

Temperature,°C	1300-1500	1200-1300	1000-1200	900-1000	750-900	700	640	525
Holding time, h	2	1	2	6	48-72	168	336	504

The segment of the phase diagram investigated is shown in Fig. 25. The eutectic reaction $Ti_5Ge_3 + \beta \rightleftarrows L$ occurs at 1410°C and 16 at-%. Germanium raises the polymorphic-transformation temperature of titanium from 872-880 to 897°C. The latter temperature corresponds to the peritectic reaction

$$\beta + Ti_5Ge_3 \rightleftarrows \alpha.$$

The solubility of germanium in α- and β-Ti decreases as the temperature falls. Germanium forms solid substitution solutions with both titanium modifications.

The β-phase does not persist at low temperatures, even after very abrupt quenching. Abruptly quenched alloys acquire a martensitic structure. Microscopic investigation of titanium-rich alloys (prepared from magnesiothermal titanium) showed that at least 1% (0.65 at-%) Ge dissolves in α-Ti at 790°C and in β-Ti at 925°C. The fact that an alloy containing 0.24% (0.16 at-%) Ge is monophasic after annealing at 790°C is in conformity with these data [209].

The chemical compound Ti_5Ge_3 has a hexagonal structure of Mn_5Si_3 type with the lattice constants a = 0.7537, c = 0.5223 nm, and c/a = 0.693 [105, 106].

The compound $TiGe_2$ has a rhombic structure of the $TiSi_2$ type with the lattice constants a = 0.8577, b = 0.5020, and c = 0.8846 nm [82]. According to Kornilov [212], $TiGe_2$ is formed at a temperature of 897°C.

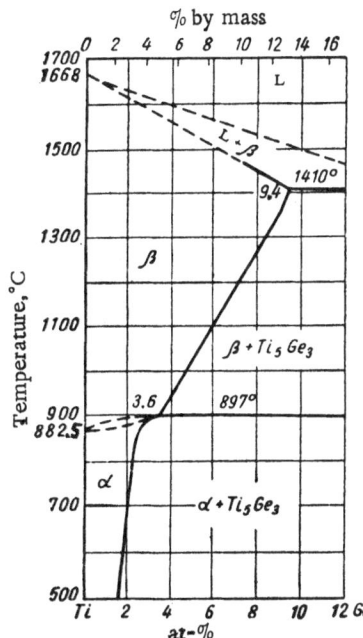

Fig. 25. Section of the phase diagram of the titanium-germanium system [208].

Ageev and Samsonov [211] investigated the compound TiGe in monocrystals taken from shrink holes in an ingot. The calculated lattice constants were a = 0.3807, b = 0.5230, and c = 0.6833 nm. They confirmed the validity of the assumption that TiGe is isostructural with TiSi.

In addition to the compounds described above, Rossteutscher and Schubert [112] showed that this system contains the compound Ti_3Ge, which has a structure of the Fe_3P type and the lattice constants a = 1.029, c = 0.514 nm, and c/a = 0.50.

Ageev et al. [210] studied the character of the interatomic interaction in Ti_5Ge_3, TiGe, and $TiGe_2$, proceeding from the decrease in interatomic distance in these structures. The trend of the variation in minimum interatomic distance in titanium germanides is, in general, similar to that observed for titanium silicides. The most stable of these germanides is TiGe, for which the deviation of the experimentally determined minimum length of the Ti−Ge bond from the additively calculated value is the most negative (−0.0250 nm); it is followed by Ti_5Ge_3 (−0.0110 nm) and $TiGe_2$ (+0.0050 nm). The Ti−Ti distance observed in α-Ti was used to calculate the additive Ti−Ge distances, as for the Ti−Si system. Heteroatomic bonds have the highest values in titanium germanides.

The gram-atomic volumes of the elementary titanium and germanium in the titanium−germanium system are 10.7 and 13.6 cm^3, which corresponded to increments of 5.2 cm^3 for titanium and 16.3 cm^3 for germanium; on this basis, we obtain changes of 0.486 and 1.20 respectively on switching from atomic volume to increment. The titanium atoms are thus compressed in the alloy, while the germanium atoms expand, this corresponding to electron transfer from the titanium to the germanium. The titanium carries an excess positive charge and the germanium a negative charge in the alloys of this system [210].

Addition of germanium markedly increases the Vickers hardness of titanium. The change in the hardness of cast titanium-rich alloys with composition is represented by the graph in Fig. 26. Addition of less than 0.24% by mass germanium somewhat increases the yield strength and relative reduction in area of annealed hot-rolled titanium and reduces its ultimate strength.

The change in the mechanical properties of titanium containing small amounts of germanium during tensile testing is characterized by the curves in Fig. 27. The presence of less than 0.3% Ge somewhat improves the machinability of titanium [209].

The chemical compounds Ti_5Ge_3 and $TiGe_2$ do not exhibit superconductivity until a temperature somewhat below 1.2°K is reached [109, 110]. Superconductivity at lower temperatures has not been determined.

Zirconium Germanides. Wallbaum [82] reported the existence of the intermetallic compound $ZrGe_2$, which has a rhombic structure of the $ZrSi_2$ type and the constants a = 0.3804, b = 1.501, and c = 0.3764 nm.

A rough phase diagram of the germanium−zirconium system was constructed on the basis of the data of Lustman and Kerze [213] (Fig. 28). It presumed the existence of the chemical compound Zr_3Ge (20.97% Ge), Zr_2Ge (28.47% Ge), and $ZrGe_2$. In order to construct a complete phase diagram (Fig. 29), Carlson [214] employed thermal, micro-

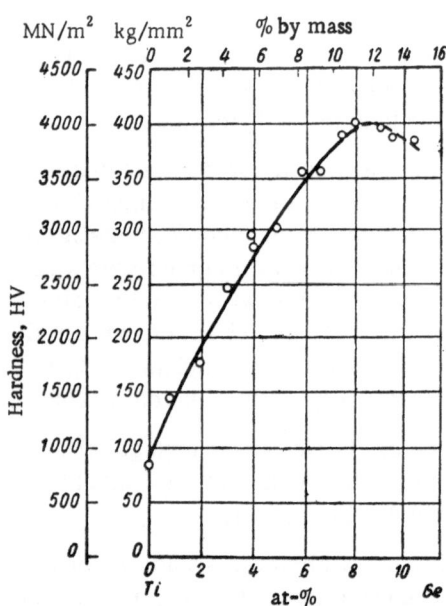

Fig. 26. Change in hardness of titanium-rich Ge−Ti alloys with composition [208].

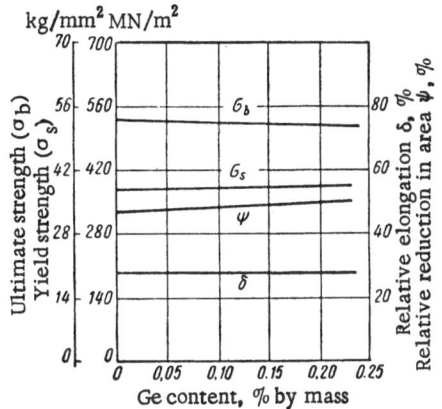

Fig. 27. Influence of addition of germanium on mechanical properties of titanium [209].

scopic, and x-ray analysis and determined the melting-initiation point at temperatures above 2000°C. The alloys were prepared from zirconium iodide and powdered germanium (99.9% pure), which were preliminarily treated with hydrogen at 650°C to reduce the GeO$_2$ and refused in a graphite crucible under a vacuum of less then 13.33 mN/m^2 (10^{-4} mm Hg). Alloys containing up to 80% Ge were prepared by fusion in a water-cooled crucible in a helium atmosphere in an arc furnace with a nonreactive tungsten electrode. Alloys containing more than 80% Ge were fused in a graphite crucible in a vacuum.

The diagram in Fig. 29 differs from that previously constructed in the following ways:

1. The maximum solubility of germanium in solid zirconium at 1535°C is 0.5-1%.

2. The compound Zr$_2$Ge is replaced by Zr$_5$Ge$_3$ (32.2% Ge), which has a melting point of 2330°C, a hexagonal structure of the Mn$_5$Si$_3$ type, and the lattice constants a = 0.799, c = 0.554 nm, and c/a = 0.693 at 31.2% or a = 0.7993, c = 0.5597 nm, and c/a = 0.700 [87]. Its experimental density is 7.12 g/cm^3 and its x-ray density is 7.29 g/cm^3.

3. There is a compound with the nominal formula ZrGe, which is formed by a peritectic reaction at 2240°C. Schubert et al. [426] determined its structure and composition (Table 1).

The compound ZrGe is quite brittle.

4. The compounds Zr$_3$Ge and ZrGe$_2$ are also formed by peritectic reactions, at temperatures of 1585 and 1520°C respectively.

The structure of ZrGe$_2$, which had previously been established by analysis of powder x-ray patterns [82], was confirmed by investigation of monocrystals [215]. Precise measurement of the lattice constants yielded the values a = 0.3789, b = 1.4957, and c = 0.3760 nm.

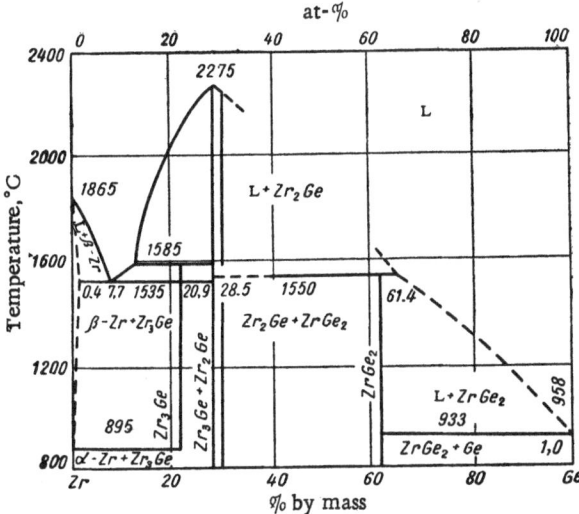

Fig. 28. Rough phase diagram of the zirconium—germanium system [213].

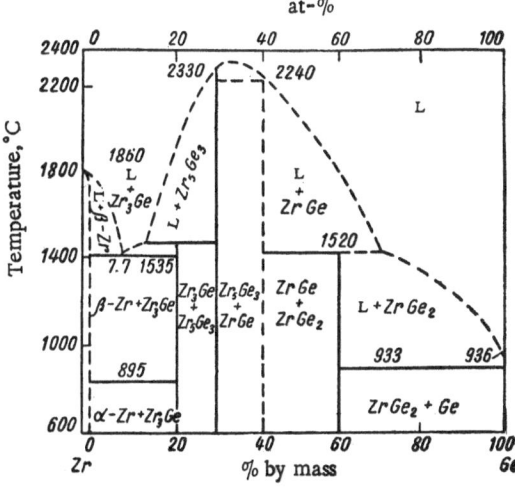

Fig. 29. Phase diagram of the zirconium—germanium system [214].

The compound ZrGe has a hexagonal lattice with the constants a = 0.814, c = 0.717 nm, and c/a = 0.880; the elementary cell contains five formal units [214].

According to other data [112], Zr_3Ge has a tetragonal structure of the Ti_3P type with the lattice constants a = 1.108, c = 0.548 nm, and c/a = 0.49. It was also established that the compound Zr_5Ge_4, which has a Zr_5Si_4 structure and the lattice constants a = 0.724, c = 1.316 nm, and c/a = 1.81, exists in addition to the compounds described above.

When Zr−Ge alloys were investigated at temperatures down to 1.2°K [109, 110], they did not exhibit superconductivity and the chemical compound $ZrGe_2$ was not detected.

Hafnium Germanides. The hafnium−germanium system has been found to contain the compounds $HfGe_2$ (44.84% Ge), which has a rhombic structure of the $ZrSi_2$ type and the lattice constants a = 0.3815, b = 1.500, and c = 0.3779 mm [215], and Hf_5Ge_3, which has a structure of the $D8_8$ type [108]. Nowotny et al. [86] confirmed the existence of the aforementioned compounds and established that of the germanide Hf_2Ge, which is isostructural with $CuAl_2$ and has the lattice constants a = 0.6574, c = 0.5361 nm, and c/a = 0.8156.

X-ray diffraction analysis of alloys containing principally the $D8_8$ phase yielded different lattice-constant values, ranging from a = 0.7855, c = 0.5546 nm, and c/a = 0.706 to a = 0.7895, c = 0.5534 nm, and c/a = 0.701.

A small amount of the monogermanide HfGe, with a lattice of the FeB type, was found in addition to the $D8_8$ phase and $HfGe_2$ in specimens containing 50 at-% Ge. The lattice constants were not calculated, since the lines were few and weak, but they are close to those for the isostructural compound ZrSi.

According to Parthe [108], the x-ray patterns of specimens corresponding to the composition Hf_5Ge_3 with 3 at-% C added indicate a hexagonal structure with the lattice constants a = 0.7883, c =5537 nm, and c/a = 0.702.

In addition to the aforementioned compounds, Rossteutscher and Schubert [112] reported that this system contains the compounds Hf_3Ge, which has a structure of the Ti_3P type and the lattice constants a = 1.092, c =0.542 nm, and c/a = 0.49, and Hf_3Ge_2, which has a structure of the U_3Si_2 type and the lattice constants a = 0.706, c = 0.372 nm, and c/a = 0.526.

GERMANIDES OF SUBGROUP-Va METALS

Vanadium Germanides. The germanium-vanadium system was previously found to contain the compound V_3Ge (32.2% Ge), which has a structure of the β-W type with a = 0.4759 nm [82].

This system was studied most completely by Holleck et al. [111], who confirmed the existence of V_3Ge and established that of the new compounds V_5Ge_3 (with a structure of the $D8_8$ type), V_3Ge_2, and VGe_2.

These authors found that the solubility of vanadium in germanium is negligibly small and that the lattice constant of the solid solution differs only slightly from that of vanadium.

No intermediate phases were detected in the composition region between pure vanadium and the compound V_3Ge.

Two phases, with structures of the T1 and $D8_8$ types, were found in addition to V_3Ge in alloys containing 30 at-% Ge. The main phase in alloys containing 38 at-% Ge was V_5Ge_3, which is of the T1 type and has the lattice constants a = 0.955, c = 0.483 nm, and c/a = 0.506.

Specimens produced by hot pressing contained a V_5Ge_3 phase of the $D8_8$ type with a broad homogeneity region. Table 4 gives the lattice constant of this compound as a function of composition.

Table 4. Lattice Constants of $D8_8$ Phase

Ge content of specimen, at-%	Lattice constants, nm		c/a
	a	c	
38	0.7262	0.4953	0.682
40	0.7273	0.4958	0.682
75	0.7250	0.4967	0.685

The compound V_3Ge_2, which is isostructural with Cr_3Ge_2, was found in alloys containing 45 at-% Ge produced by cold pressing and sintering. This compound was not detected in specimens produced by hot pressing.

A new phase appeared in the x-ray diffraction patterns of specimens containing 55 at-% Ge and was especially distinct at 67 at-% Ge. The authors assume that this phase is the compound VGe_2, which is formed very slowly and therefore could not be produced in sufficiently homogeneous form. Its formation is hampered by the development of the $D8_8$ phase.

X-ray diffraction analysis of monocrystals with the approximate compositions V_3Ge_2 and $VGe_{1.8}$ [216] made it possible to establish the structure and composition of these compounds ($V_{11}Ge_8$ and $VGe_{1.8}$). According to later data [217], the latter compound should be assigned the formula $V_{17}Ge_{31}$.

The compound $V_{11}Ge_8$ has a structure of the Mn_5Si_3 type with the constants a = 1.341, b = 1.609, and c = 0.502 nm. Its pycnometric density is 6.89 g/cm^3 and its x-ray density is 6.99 g/cm^3.

The compound $V_{17}G_{31}$ ($VGe_{1.8}$) has a complex lattice made up of simple sublattices of vanadium atoms; its structure is a modified $TiSi_2$ type and has the constants a = 0.590, c = 8.367 nm (a' = 0.590, c' = 0.492 nm, and c'/a' = 0.8339).

Sarachik et al. [218] measured the thermal emf of V_3Ge [as well as V_3Si, V_3Ga, and V_3Sn] at temperatures ranging from the transition point to the superconductive state to room temperature. They established that the temperature coefficients of thermal emf are positive for all the compounds in question, equalling + 10 μV/deg at room temperature. Investigation of the coefficient of thermal emf as a function of temperature (Fig. 30) showed that it increases monotonically. According to the data obtained by these authors, the resistance of V_3Ge at room temperature is ρ = 89 μohm·cm, its electronic heat capacity is γT = 9.2 J/deg·mole (2.2 cal/deg·mole), and its thermal emf is α = 31 μV/deg. A thin film of VGe does not exhibit superconductivity at temperatures above 1.20°K [109, 110].

Niobium Germanides. Carpenter and Searcy [113] employed x-ray diffraction analysis of alloys produced by heating a mixture of 99.4% powdered niobium and 99.9% powdered germanium (both with a grain size of 200μ) in a vacuum to establish the existence of germanides having the compositions Nb_3Ge (20.66% Ge), $NbGe_{0.67 \pm 0.06}$, ~Nb_3Ge_2 (34.25% Ge), and $NbGe_{0.54 \pm 0.06}$, with $NbGe_{0.6}$ corresponding to the formula Nb_3Ge_3 (31.88% Ge). Wallbaum [82] and Hardy and Hulm [109, 110], who employed specimens prepared by a powder-metallurgical method and by direct fusion, reported the compound $NbGe_2$ and Nb_2Ge.

The compound Nb_3Ge is stable below 1910°C and has a structure of the β-W type with the lattice constant a = 0.5168 nm (O_h^3-Pm3n space group). This compound is formed by heating $NbGe_{0.54}$ and Nb at 1700°C for more than 3 h, in order to establish equilibrium.

Wallbaum [82] reported that $NbGe_2$ has a structure of the $CrSi_2$ type with the lattice constants a = 0.4957, c = 0.6770 nm, and c/a = 1.366. Carpenter and Searcy [113] found that this

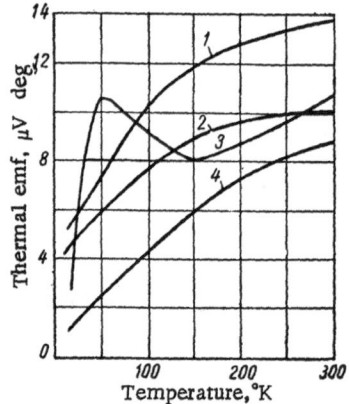

Fig. 30. Thermal emf of compounds of R_3X type as a function of temperature [218]. 1) V_3Si; 2) V_3Ga; 3) V_3Ge; 4) V_3Sn.

compound has a hexagonal lattice with the constants a = 0.4966, c = 0.6781 nm, and c/a = 1.365. It decomposes at a temperature of 1483°C.

According to the most recent data [72], the compound $NbGe_2$ has a lattice of the $CrSi_2$ type with the constants a = 0.4943, c = 0.6778 nm, and c/a = 1.371 and three molecules in the elementary cell. It has a pycnometric density of 8.17 g/cm^3 and an x-ray density of 8.295 g/cm^3. These authors also determined certain of its physical properties. Thus, its resistivity is 67 and 31 μohm·cm at 25 and 196°C respectively, its coefficient of thermal conductivity is 0.31 W/cm·deg at 25°C, and its temperature coefficient of thermal emf is + 12 μV/deg.

Nowotny et al. [219] studied the crystal structure of $NbGe_{0.54 \pm 0.06}$ and $NbGe_{0.67 \pm 0.05}$.

The compound $NbGe_{0.67 \pm 0.05}$ has a hexagonal structure of the Mn_5Si_3 type with the lattice constants a = 0.7718, c = 0.5370 nm, and c/a = 0.6958.

The compound $NbGe_{0.54}$, which is homogeneous over the composition range 32.4–37.5 at-% Ge, is isostructural with Mo_5Si_3 and has a tetragonal lattice with the constants a = 1.0148, c = 0.5152 nm, and c/a = 0.508.

The compounds $NbGe_{0.67}$, $NbGe_{0.54}$, and Nb_3Ge do not decompose at temperatures below 1646, 1938, and 1938°C respectively. Alloys with compositions lying between $NbGe_{0.54}$ and Nb_3Ge melt at 1910°C. The pycnometric densities of Nb_3Ge and $NbGe_2$ are 8.17 and 7.81 g/cm^3 respectively, while their x-ray densities are 8.47 and 8.20 g/cm^3 [113]. Pan et al. [495] worked out the phase diagram of the niobium−germanium system (Fig. 30a) by microscopic, thermal, and x-ray analysis.

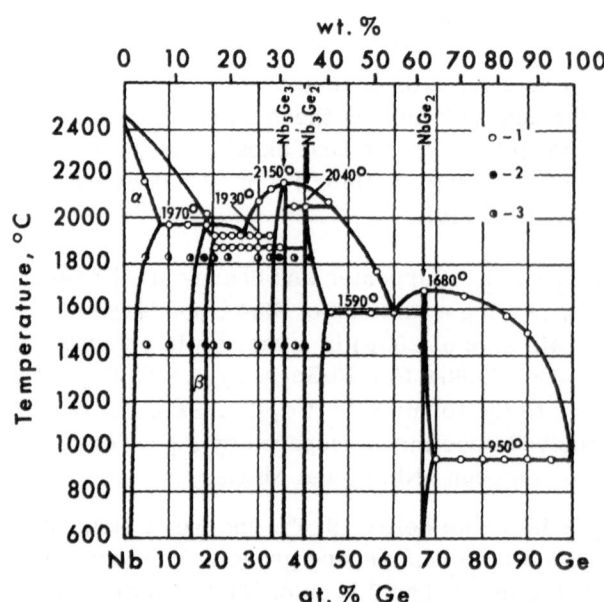

Fig. 30a. Phase diagram of the niobium−germanium system [495].

The solid solution of germanium in niobium forms a narrow monophasic region; the maximum solubility of germanium in niobium is about 7 at-% at the peritectic equilibrium temperature (1970°C), while the solubility of niobium in germanium is extremely low.

The β-phase, which is based on the compound Nb_3Ge, is formed by the peritectic reaction $\alpha + L \rightarrow \beta-Nb_3Ge$ at 1970°C. The homogeneity region of this phase is displaced toward niobium from its stoichiometric composition and lies at 16-21 at-% Ge for 1850°C and at 15-19 at-% Ge for 1450°C.

X-ray analysis of cast niobium—germanium alloys annealed at 1850 and 1450°C showed both $\beta-Nb_3Ge$ and three intermediate phases, which lay at 34-36 and 40-43 at-% Ge (the authors arbitrarily designated them as Nb_5Ge_3 and Nb_3Ge_2 at 66.6 at-% Ge ($NbGe_2$).

The compound Nb_5Ge_3 melts congruently at 2150°C. The eutectic equilibrium $L \rightleftharpoons Nb_3Ge + Nb_5Ge_3$ occurs at a temperature of 1930°C and a Ge content of 26-27 at-%.

The compound Nb_3Ge_2 has a monophasic region at Ge contents of 40-43 at-% and is formed by the peritectic reaction $Nb_5Ge_2 + L \rightarrow Nb_3Ge_2$ at 2040°C.

The compound $NbGe_2$ melts congruently at 1680°C and forms a eutectic with the Nb_3Ge_2 at 1590°C and with the germanium at 950°C.

The compound Nb_2Ge passes into the superconductive state at 1.2°K [109, 110], while according to Roberts [12], the germanides $NbGe_{0.22}$, $Nb_{3.45}Ge_{0.55}$, and $Nb_{3.28}Ge_{0.72}$ become superconductive at temperatures of 5.3, 4.9, and 5.5°K respectively. However, according to the same data, the compound Nb_5Ge_3 does not exhibit superconductivity at temperatures above 1.1°K.

Carpenter et al. [114, 115] employed x-ray analysis and Knudsen's effusion method to study the homogeneity region, decomposition pressure, and thermodynamic stability of the compounds Nb_3Ge, α- and $\beta-Nb_5Ge_3$, and $NbGe_2$.

The x-ray data established that Nb_3Ge has a homogeneity region at 1600°C, extending from $NbGe_{0.22 \pm 0.02}$ to $NbGe_{0.15 \pm 0.01}$ with lattice constants of a = 0.5167 nm to a = 0.5177 nm respectively [114].

The compound Nb_3Ge decomposes by the reaction

$$6.67\ NbGe_{0.15(S)} = 6.67\ Nb_{(S)} + Ge_{(G)}$$

and the equilibrium decomposition pressure varies from 13.04 mN/m^2 (1.33·10^{-7} atm) to 135.3 mN/m^2 (1.38·10^{-6} atm) over the temperature range 1610-1752°C. The enthalpy of decomposition for the reaction is $\Delta H°_{298}$ = 448 kJ/mole (116.6 kcal/mole). If we assume the enthalpy of sublimation of germanium to be 376.81 kJ/mole (90 kcal/mole), the enthalpy of formation of Nb_3Ge by the reaction $1/0.15\ Nb_{(S)} + Ge_{(S)} = 1/0.15\ NbGe_{0.15(S)}$ is $\Delta H°_{298}$ = -111.3 kJ/mole (-26.6 kcal/mole) [114].

The enthalpy of decomposition of $\alpha-Nb_5Ge_3$ by the reaction

$$3.12\ NbGe_{0.51(S)} = 3.12\ NbGe_{0.22(S)} + Ge_{(G)}$$

is 512.04 kJ/mole (122.3 kcal/mole) [115].

The equilibrium decomposition pressure for $\beta-Nb_5Ge_3$ in the reaction 7.69 $NbGe_{0.67(S)}$ = 7.69 $NbGe_{0.54(S)}$ + $Ge_{(G)}$ is 2.393 N/m^2 (2.44·10^{-5} atm) at 1484°C.

The enthalpies of formation of the compounds $\alpha-Nb_5Ge_3$, $\beta-Nb_5Ge_3$, and $NbGe_2$ are -113.4 kJ/mole (-27.1 kcal/mole), -95.8 kJ/mole (-22.9 kcal/mole), and -43.5 kJ/mole (-10.4 kcal/mole) respectively [115].

   Tantalum Germanides. Wallbaum [82] reported the first data on the structure of
the compound $TaGe_2$ (44.51% Ge). This compound has a hexagonal structure of the $CrSi_2$ type
with the lattice constants a = 0.4948, c = 0.6737 nm, and c/a = 1.362.

   These data are in good agreement with those of Brixner [72], who showed that $TaGe_2$ has
a lattice of the $GrSi_2$ type with the constants a = 0.4943, c = 0.6737 nm, and c/a = 1.363 and three
molecules in the elementary cell. This compound has a pycnometric density of 11.28 g/cm^3
and an x-ray density of 11.392 g/cm^3.

   The resistivity of $TaGe_2$ is 35 and 13 μohm·cm at 25 and –196°C respectively, its coefficient of thermal conductivity is 0.35 W/cm·deg at 25°C, and its temperature coefficient of thermal emf is 11 μV/deg.

   In addition to the digermanide, Hardy and Hulm [109, 110] were able to observe an intermediate phase with the approximate composition Ta_2Ge, which is produced by decomposition of
$TaGe_2$ at 1280°C.

   The chemical compound $TaGe_2$ does not exhibit superconductivity until the temperature
is reduced to below 1.2°K [109, 110].

   Nowotny [219] discovered the compound Ta_5Ge_3, which exists in two modifications:
1) the α-modification of Ta_5Ge_3 corresponds to the high–temperature modification of Ta_5Si_3 and
has a tetragonal lattice with the constants a = 0.6599, c = 1.2010 nm, and c/a = 1.820; 2) the
β-modification of Ta_5Ge_3 has a tetragonal structure of the Mo_5Si_3 type with the lattice constants
a = 1.0010, c = 0.5150 nm, and c/a = 0.514.

   Parthe and Norton [87] confirmed the existence of the compound Ta_5Ge_3. X-ray diffraction analysis of specimens prepared by hot pressing of tantalum hydride and germanium powders showed the binary phase Ta_5Ge_3 to have a T1 structure. When 5 at-% C was added, however, the powder pattern indicated a Nowotny phase with a $D8_8$ structure (D_{6h}^3-C6/mcm space
group) and the lattice constants a = 0.7581, c = 0.5235 nm and c/a = 0.690. The agreement between the observed and calculated intensities showed that the tantalum atoms are in the 4(d) and
6(g) positions with x = 0.25 and that the germanium atoms are in the $6(g)_{II}$ positions with x = 0.61.

   According to Rossteutscher and Schubert [112], this system contains both Ta_2Ge and the
compound Ta_3Ge, the latter having high-temperature and low-temperature modifications. The
high-temperature modification has an Fe_3P structure with the lattice constants a = 1.036,
c = 0.516 nm, and c/a = 0.50; the low-temperature modification has a Ti_3P structure with the
lattice constants a = 1.028, c = 0.522 nm, and c/a = 0.51.

GERMANIDES OF SUBGROUP-VIa METALS

   Chromium Germanides. X-ray study of the chromium−germanium system [82]
has demonstrated the existence of three compounds: Cr_3Ge (31.76% Ge) has a lattice of the
β-W type with the constant a = 0.4614 nm and is a superstructure isomorphic with Cr_3Si;
Cr_3Ge_2 (48.31% Ge) has a rhombic structure of the Cr_3Si_2 type; CrGe (58.26% Ge) has a cubic
structure of the FeSi type and the lattice constant a = 0.4780 nm. According to the data of Zagryazhskii et al. [220], the compound Cr_3Ge has a homogeneity region extending from 23.1-25.7 at-%
Ge at 1150°C and melts congruently at 1520-1540°C [85].

   In addition to the previously known compounds $CrGe_2$ and Cr_3Ge_2, Voellenkle et al. [216]
reported the existence of $Cr_{11}Ge_{19}$ and $Cr_{11}Ge_8$. The alloys were prepared by fusing mixtures of
metallic chromium and germanium powders in a quartz crucible in an argon atmosphere in a
high-frequency furnace. After annealing at 800°C for 100 h in an evacuated and sealed quartz
tube, the alloy containing 66.67 at-% Ge was homogeneous and identical to the phase $CrGe_2$. Since
free germanium was deposited on the walls of the tube during annealing in all cases, it was

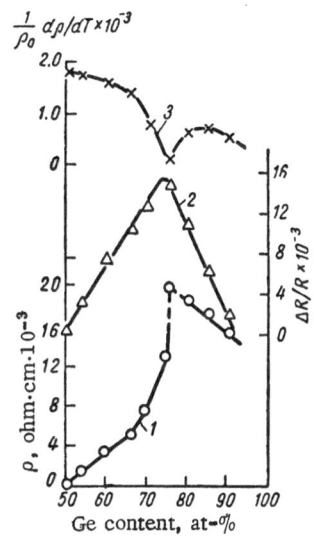

Fig. 31. Dependence of ρ (1), $\Delta R/R$ (2), and $1/\rho \cdot d\rho/dt$ (3) on composition in Cr−Ge alloys [117].

concluded that the resultant compound had a composition poorer in germanium that $CrGe_2$. It was assigned the formula $Cr_{11}Ge_{19}$. X-ray analysis of monocrystals of this compound showed it to have a tetragonal cell of the $Mn_{11}Si_{19}$ type with the constants a = 0.580 and c = 5.234 nm (a' = 0.580, c' = 0.476 nm, and c'/a' = 0.8207).

The compound $Cr_{11}Ge_8$ has a structure of Mn_5Si_3 type with the constants a = 1.315, b = 1.575, and c = 0.495 nm. Its pycnometric density is 7.47 g/cm^3 and its x-ray density is 7.48 g/cm^3.

Parthe and Norton [87] established the existence of the chemical compound Cr_5Ge_3 (45.58% Ge by mass), which has a structure of the Cr_5Si_3 type with the lattice constants a = 0.9413, c = 0.4780 nm, and c/a = 0.507 and four molecules in the elementary cell.

Fakidov and Grazhdankina [117], who investigated alloys in the germanium−chromium system, established the existence of the compound $CrGe_3$, which exhibits ferrromagnetism.

Table 5 and Fig. 31 present the results of measurements of the resistivity ρ, mean temperature coefficient of electrical resistance $\bar{\beta}$, change in resistance in a magnetic field $\Delta R/R$, and Curie temperature of chromium−germanium alloys with different chemical compositions.

As can be seen from the data obtained, the resistivity of chromium−germanium alloys is 2–4 orders of magnitude greater than that of ordinary metals and has a maximum at 75 at-% Ge.

The values of $\bar{\beta}$ for alloys containing up to 70 at-% Ge are of the same order of magnitude as for ordinary metals. There is a decrease in $\bar{\beta}$ with increasing germanium content.

Research has shown that alloys with a high germanium content (50% or more) become ferromagnetic when cooled to the temperature of liquid nitrogen. On the basis of Curie-temperature measurements, Fakidov and Grazhdankina [117] found that the ferromagnetism in this system is due to the presence of only one ferromagnetic phase. Considering the extrema of the curves representing ρ, $\Delta R/R$, and $1/\rho_0 \cdot d\rho/dt$ as a function of alloy composition in the vicinity of a Ge concentration of 75 at-%, they surmised that the ferromagnetic phase must be the chemical compound $CrGe_3$ or solid solutions based on this compound.

TABLE 5. Electromagnetic Properties of Ge−Cr Alloys

Alloy composition				$\rho \cdot 10^{-3}$ ohm·cm	$\bar{\beta} \cdot 10^{-3}$	$\Delta R/R \cdot 10^{-3}$ (H= 10,000 Oe)	θ_f, °K
at-%		% by mass					
Cr	Ge	Cr	Ge				
50.0	50.0	42.7	57.3	0.19	1.80	0	98.0
46.0	54.0	38.0	62.0	1.52	1.75	2.9	98.0
40.0	60.0	32.3	67.7	3.50	1.54	−7.4	98.0
34.0	66.0	27.0	73.0	5.02	2.37	−10.1	98.0
30.0	70.0	23.5	76.5	7.47	0.71	−12.5	98.0
25.0	75.0	19.3	80.7	13.20	0.80	−15.0	98.4
20.0	80.0	15.2	84.8	18.42	0.82	−10.6	98.8
15.0	85.0	11.2	88.8	17.00	0.72	−6.3	98.8
10.0	90.0	7.4	92.6	15.32	0.58	−1.9	98.6
5.0	95.0	3.6	96.4	13.9	0.28	—	98.7

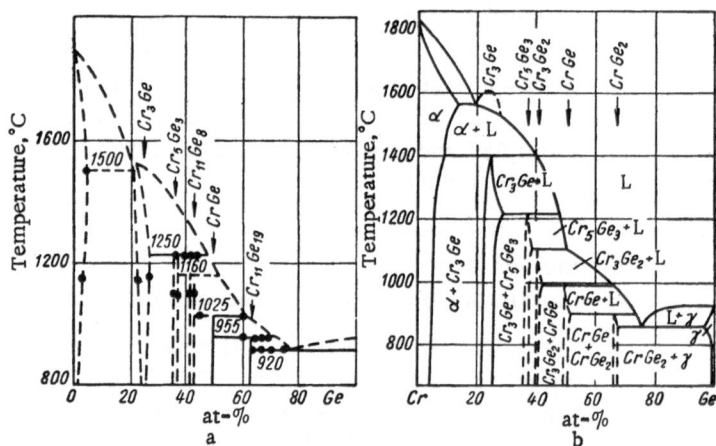

Fig. 32. Phase diagrams of the chromium−germanium
system [442, 449].

However, Margolin and Fakidov [126], who studied the magnetic properties of Cr−Ge alloys, noted that maximum specimen magnetization occurred for $CrGe_2$. These authors did not make metallographic or x-ray analyses of the phase composition of the alloys, which casts doubt on the existence of a compound with this composition. According to the data of Voellenkle [216], the highest chromium germanide is assigned the formula $Cr_{11}Ge_{19}$.

The chromium−germanium system has been most completely studied by Zagryazhskii et al. [442, 445], who proposed a phase diagram for this system (Fig. 32a). As can be seen from this diagram, the system contains the compound Cr_3Ge, Cr_5Ge_3, $Cr_{11}Ge_8$, $CrGe$, and $Cr_{11}Ge_{19}$; all these, with the exception of Cr_3Ge, are formed peritectically.

The maximum solubility of germanium in chromium at 1150°C is about 3 at-%.

The compound $Cr_{11}Ge_{19}$ is formed by the peritectic reaction $CrGe + L \rightleftharpoons Cr_{11}Ge_{19}$ at 955°C and produces a eutectic with germanium that melts at 920°C.

Microscopic and thermal analysis has shown that chromium monogermanide, $CrGe$, is formed peritectically at 1025°C by the reaction $Cr_{11}Ge_8 + L \rightleftharpoons CrGe$, while $Cr_{11}Ge_8$ is formed by the reaction $Cr_5Ge_3 + L \rightleftharpoons Cr_{11}Ge_8$ at 1160°C. The pycnometric density of the latter compound is 7.40 g/cm^3.

Zagryazhskii [445] noted that the germanide Cr_5Ge_3 is a phase of variable composition. It has been shown by pycnometric and x-ray investigations that a solid substitution solution is formed, the parameters of the tetragonal Cr_5Ge_3 cell being independent of composition. This compound is formed by the peritectic reaction $Cr_3Ge + L \rightleftharpoons Cr_5Ge_3$ at about 1250°C. The compound Cr_3Ge melts congruently; the alloy containing 22 at-% Ge has the highest melting point.

Svechnikov and Kobzenko [449] proposed a somewhat modified phase diagram for the germanium−chromium system (Fig. 32b). It is distinguished by simpler atomic ratios for the germanides (Cr_3Ge_2 and $CrGe_2$), by the presence of a monotectic, and by the absence of congruently melting germanides.

Zagryazhskii et al. [486] measured certain of the electrophysical properties of the germanide $Cr_{11}Ge_{19}$. The results of magnetic-susceptibility measurements made over the temperature range 60–950°K confirmed its complex magnetic structure. It was established that the paramagnetic Curie point is $\theta_p = 55°K$, the magnetic-disordering temperature is $\theta_c = 86°K$, and the mean magnetic moment of the chromium atoms is $\mu_{eff} = 2.55 \ \mu V$ (which evidently indicates the presence of two magnetic electrons and a tetravalent atomic state).

The data given above differ materially from the results obtained by Margolin and Faki-dov [126], according to whom θ_p = 142°K, θ_k = 100-110°K, and μ_{eff} = 2.3 μV. These discrepancies apparently due to the fact that nonequilibrium specimens differing in composition from $Cr_{11}Ge_{19}$ were used.

The compound $Cr_{11}Ge_{19}$ exhibits conductivity of the metallic type over the temperature 80-1000°K. Its resistivity at room temperature is 3.45·10^{-4} ohm·cm. As with other chromium germanides (Cr_5Ge_3 and CrGe), the temperature coefficient of resistance for $Cr_{11}Ge_{19}$ decreases monotonically when the temperature is raised above 100°K. Its thermal emf varies with temperature in similar fashion, increasing smoothly from + 10 μV/deg at 80°K to 32 μV/deg at 300°K. Judging from the sign of α and the value of the Hall constant R (R = + 1.15·10^{-3} cm^3/C at 300°K), holes are the principal type of charge carrier in $Cr_{11}Ge_{19}$, as in the highest chromium silicides.

None of the chemical compounds of germanium with chromium exhibit superconductivity at low temperatures [109, 110].

Molybdenum Germanides. Searcy and Peavler [118] employed the x-ray method to establish the existence of the phases Mo_3Ge_2 (33.53% Ge), Mo_2Ge_3 (53.16% Ge), and $MoGe_2$ (60.20% Ge) in the germanium−molybdenum system, in addition to the previously discovered phase Mo_3Ge (20.14% Ge) [221].

Two polymorphic modifications were found for $MoGe_2$: α-$MoGe_2$ is the low-temperature form and β-$MoGe_2$ is the high-temperature form. The latter is isostructural with $MoSi_2$ and has a tetragonal structure with six atoms in the elementary cell and the lattice constants a = 0.3313, c = 0.8195 nm, and c/a = 2.474 [118]. It belongs to the D_{4h}^7-14/mmm space group.

The specimens for x-ray diffraction analysis were prepared by heating a mixture of germanium and molybdenum powders with a grain size of 200 μ in graphite crucibles.

The calculated density of β-$MoGe_2$ is 8.91 g/cm^3, with two molecules in the elementary cell [118]. An interaction between the elements in pressed powder mixtures was observed at 722°C [223].

Stecher et al. [224] have conducted the most complete investigation of the molybdenum−germanium system. They employed melting-point data and microscopic and x-ray analysis to construct the phase diagram of this system (Fig. 33).

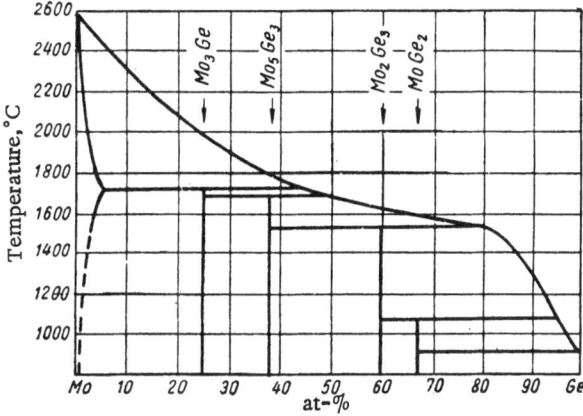

Fig. 33. Phase diagram of the molybdenum−germanium system [224].

The maximum solubility of germanium in molybdenum is 5 at-% at 1750°C and drops as the temperature is reduced. The solubility of molybdenum in germanium is negligible.

The phases Mo_3Ge, Mo_5Ge_3, Mo_2Ge_3, and $MoGe_2$ have been found to exist in this system, being formed peritectically.

The compound Mo_3Ge is formed from the molybdenum and the melt at 1750°C.

The phase Mo_5Ge_3 (Mo_3Ge_2) is formed from the melt and Mo_3Ge at 1730°C. It has a lattice of the W_5Si_3 type with the constants a = 0.9837, c = 0.4973 nm, and c/a = 0.506.

The x-ray patterns of specimens containing 68 at-% Ge (1350°C) exhibit lines for a compound with a structure of the $TiSi_2$ type, in addition to germanium lines. This phase is rather homogeneous at 60 at-% Ge, which corresponds to Mo_2Ge_3. The compound is formed from Mo_5Ge_3 and the melt at 1520°C, is stable, and has a rather broad homogeneity region.

However, it has been found [216] that a compound with the composition $Mo_{13}Ge_{23}$ exists in addition to Mo_2Ge_3. In order to obtain the compound described in the literature as Mo_2Ge_3, mixtures of molybdenum and germanium powders with different compositions were annealed in an evacuated and sealed quartz ampule at 1000°C for 1 h. X-ray diffraction analysis established that the readily sintered product with the composition $MoGe_{1.8}$ was virtually homogeneous. Annealing in evacuated quartz tubes at 1200°C for 20 h yielded single crystals of this compound, which were subjected to x-ray analysis. These investigations showed that the compound has a tetragonal lattice (D_{2d}^8–P4n2 space group) of complex-lattice type that is a modification of the $TiSi_2$ structure and the ideal composition $Mo_{13}Ge_{23}$. The elementary cell of $Mo_{13}Ge_{23}$ is built up of simple sublattices of molybdenum atoms and has the constants a = 0.599 and c = 6.354 nm (a' = 0.599, c' = 0.489 nm, and c'/a' = 0.816). The compound has a pycnometric density of 8.56 g/cm^3 and an x-ray density of 8.52 g/cm^3.

The compound $MoGe_2$, which is a low-temperature digermanide form with an undetermined structure, appears at temperatures below 1100°C in alloys containing more than 60 at-% Ge. No high-temperature $MoGe_2$ phase of the $MoSi_2$ type has been detected.

Brown [225] prepared the compounds $MoGe_2$ and $MoGe_{1.7}$ and determined their structure. The compound $MoGe_2$ has a rhombic structure of the $PbCl_2$ type (Pnma space group) with the lattice constants a = 0.6343, b = 0.3451, and c = 0.8582 nm.

The compound $MoGe_{1.7}$ has a tetragonal structure (14_122 space group) with the lattice constants a = 0.5994, c = 4.3995 nm, and c/a = 7.339.

Barbier-Andrieux [88, 89] reported isolation of the phases Mo_3Ge, Mo_3Ge_2, and α-$MoGe_2$ from alloys produced by electrolysis of a melt containing sodium or lithium metaborate, germanium dioxide, and molybdic anhydride.

Table 6. Physical Properties of Germanides

Compound	Resistivity ρ, μohm·cm	Hall coefficient $R \cdot 10^4$, cm^3/C	Temperature coefficient of electrical resistance $\alpha \cdot 10^3$, deg	Absolute coefficient of thermal emf α, μV/deg	Microhardness, H	
					MN/m^2	kg/mm^2
ZrGe$_2$	126	−23.8	+3.5	−	6800	680
MbGe$_2$	118	+7.0	+2.9	+12.0	7450	745
MoGe$_2$	250	−11.0	+3.2	+1.4	9400	940

The chemical compound Mo_3Ge has a cubic structure of the $\beta-W$ type with the lattice constant a = 0.4933 nm. Each germanium atom in Mo_3Ge is surrounded by 12 molybdenum atoms at a distance of 0.175 nm. Each molybdenum atom is surrounded by two germanium atoms at a distance of 0.246 nm, four germanium atoms at a distance of 0.275 nm, and eight molybdenum atoms at a distance of 0.302 nm [221]. The experimentally determined density is 9.7 g/cm^3 and the x-ray density is 9.97 g/cm^3. The heat of formation of Mo_3Ge from solid molybdenum and liquid germanium at 1527°C is $\Delta H = 60.7$ kJ/mole (14.5 kcal/mole) [226].

The dissociation pressure of this compound can be calculated from the equation: log p = $-2.14 \cdot 10^{-4}/T + 6.68$ [226].

Despite the fact that the sintering temperature was 1350°C, none of the alloys containing less than 67 at-% Ge exhibited signs of the onset of melting. The melting point of Mo_3Ge lies above 1750°C [118].

While the Mo_3Ge and Mo_3Ge_2 phases enter the superconductive state at 1.43 and 1.20°K respectively, Mo_2Ge_3 and $MoGe_2$ do not become superconductive above 1.20°K [109].

Neshpor and L'vov [68] investigated the microhardness, resistivity, Hall constant, thermal emf, and thermal coefficient of electrical resistance of zirconium, niobium, and molybdenum digermanides [$MeGe_2$]. Compact specimens of the zirconium and niobium germanides were obtained by heating the powdered metals and germanium in an argon atmosphere at 1300°C and then sintering the germanide powders by hot pressing.

Metallographic and x-ray analysis showed that the zirconium and niobium germanides were monophasic and corresponded to the compositions $ZrGe_2$ and $NbGe_2$. On the other hand, the molybdenum germanide specimens had two phases, which corresponded in composition to the two crystalline forms of $MoGe_2$.

Table 6 presents data on the physical properties of the digermanides investigated [68].

It can be seen from the data obtained that the digermanides studied have metallic conductivity, obviously of the mixed electron—hole type.

The digermanides had higher resistivities and lower microhardnesses than the corresponding disilicides [50].

According to Neshpor and Ordan'yan [227], the modulus of elasticity of $MoGe_2$ is E = 66 kN/m^2 (6.6·10^{-3} kg/mm^2).

Tungsten Germanides. Wallbaum [82] showed that the tungsten-germanium system contains a chemical compound, although neither its composition nor its structure were established. Barbier-Andrieux [88, 89] established the hypothetical existence of two chemical compounds in alloys prepared by electrolysis of a melt containing sodium metaborate, WO_3, and germanium dioxide. However, x-ray analysis of specimens produced by a powder-metallurgical method did not show the existence of chemical compounds [109, 110].

GERMANIDES OF SUBGROUP-VIIa METALS

Manganese Germanides. Zwicker et al. [119] prepared alloys by fusing electrolytic manganese and purified germanium in a hydrogen atmosphere.

Barbier-Andrieux [88, 89], who studied alloys produced by electrolysis of a melt containing sodium germanate, MnO, and sodium fluoride at 1050°C, detected lines for the compound Mn_5Ge_3 in the x-ray patterns of alloys containing from 44.9 to 45.7% Ge and lines for the compound Mn_3Ge_2 and for Ge in alloys richer in germanium. Zwicker made two studies of the phase diagram of the manganese—germanium system [119, 120]. In his first study, he employed

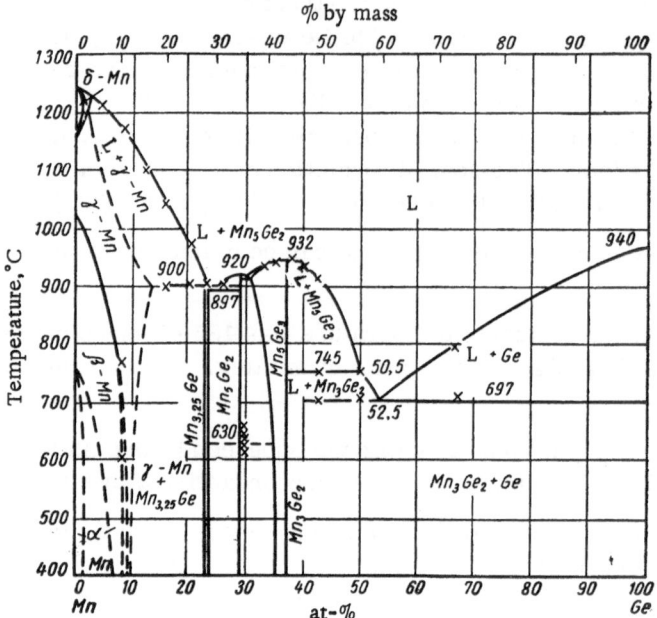

Fig. 34. Phase diagram of the manganese—germanium system [119]. X, points found by thermal and x-ray analysis.

thermal (with differential recording), microscopic, and x-ray analysis and established the existence of δ-, α-, β-, and γ-phases: $Mn_{3.25}Ge$ (28.91% Ge), Mn_5Ge_2 (34.58% Ge), Mn_5Ge_3 (44.22% Ge), Mn_3Ge_2 (46.84% Ge), and Ge. Figure 34 shows this phase diagram. The δ-, α-, β-, and γ-phases are manganese-based solid solutions.

The maxima in the phase diagram at 920 and 932°C correspond to the chemical compounds Mn_5Ge_2 and Mn_5Ge_3, which melt congruently. A broad region of solid solutions is formed on the basis of Mn_5Ge_3.

The compounds $Mn_{3.25}Ge$ and Mn_3Ge_2 are formed by peritectic transformations: Mn_3Ge_2 is formed from Mn_5Ge_3 and a melt containing 50.5 at-% Ge at 745°C, while $Mn_{3.25}Ge$ is formed from the γ-solid solution and a melt containing 23.8 at-% Ge at 900°C.

The compound Mn_3Ge_2 forms a eutectic with germanium at a temperature of 697°C and a Ge content of 59.45% (52.5 at-%). An $Mn_{3.25}Ge$—Mn_5Ge_2 eutectic is formed at a temperature of 897°C and a Ge content of 30.03% (24.5 at-%), readily decomposing when the alloy is quenched in water.

Instead of Mn_3Ge_2, Voellenkle et al. [216] established the existence of the compound $Mn_{11}Ge_3$, which has a lattice of the Mn_5Si_3 type and the constants a = 1.322, b = 1.583, and c = 0.509 nm. Its pycnometric density is 7.38 g/cm^3 and its x-ray density is 7.39 g/cm^3.

The compounds Mn_5Ge_3 and Mn_5Ge_2 also form a eutectic at a temperature of 915°C and a Ge content of 36.2% (30 at-%). In addition, they can form solid solutions, but their homogeneity region at room temperature is very small.

The compound Mn_5Ge_2 is converted to a solid at 630°C.

Zwicker et al. [120] later employed thermal and microscopic analysis to determine the polymorphic-transformation temperatures. Thus, the transformation γ-Mn $\rightleftharpoons$ β-Mn occurs at 1091°C, while the transformation β-Mn $\rightleftharpoons$ α-Mn takes place at 680°C. This same author established that addition of 1.87% Ge does not permit γ-Mn to persist on quenching from 1100°C.

Other researchers [288] have investigated the region of germanium-rich alloys (from 0 to 32 at-% Mn). These studies and those of Barbier-Andrieux [88, 89] confirmed Zwicker's data [120] indicating that the solubility of manganese in solid germanium is slight and that a eutectic exists at 720°C.

According to Zwicker et al. [119], the compound $Mn_{3.25}Ge$ has a hexagonal structure of the Mg_3Cd type with the lattice constants a = 0.5336, c = 0.4365 nm, and c/a = 0.816 in specimens containing 23.8 at-% Ge annealed at 750°C for 48 h.

At high temperatures and after quenching from 750°C, $Mn_{3.25}Ge$ has a hexagonal structure of the Mg type.

The compound Mn_5Ge_2 has a crystal structure of the NiAs type and is ferromagnetic. Yasukochi et al. [229] studied the ferromagnetic properties of this compound, which was produced by fusion of electrolytic manganese (99.9%) and germanium (99.9%). For this purpose, powdered manganese and germanium, in a ratio of 5:2, were thoroughly mixed and sealed into evacuated quartz tubes. The mixture was fused in an electric furnace at 1150°C and quenched in water at 750°C. The resultant product was then annealed at 500°C for 3 days and again quenched in water.

Investigation of the Mn_5Ge_2 obtained established that it can be regarded as a typical H-type ferromagnetic material with a Newell point at 437°C and a compensation point at 122°C.

Novogrudskii and Fakidov [230] investigated the Hall effect of Mn_5Ge_2 near its compensation point. The Hall emf E_H was measured with a low-resistance potentiometer and a galvanometer having a sensitivity of $2 \cdot 10^{-8}$ V per scale division. The Hall constant R_f was calculated from the formula $E_H = R_f MI/d$, where M is the technical magnetization of the specimen, I is the current in the specimen and d is the specimen thickness.

The R_f of Mn_5Ge_2 at a temperature $T_0 = 0.46$ is $-118 \cdot 10^{-8} \nu$ cm^{-2}/G.A. Novogrudskii and Fakidov [230] attributed this high Hall constant, similar to that previously observed for chromium telluride, to the relatively high resistivity of Mn_5Ge_2 ($\rho = 10^{-4}$ ohm·cm at 20°C).

Vrzhetsiono [232] investigated the crystal magnetic anisotropy E_C of Mn_5Ge_3 as a function of temperature T near the Curie point.

It was established on the basis of thermodynamic calculations and experimental data that this function is linear and is described by the equation

$$E_C = a_0 + C(\theta + T),$$

where a_0 and C are constants.

The compound Mn_5Ge_3 has the lattice constants a = 0.7184, c = 0.5053 nm, and c/a = 0.703 [136, 137]. It is also ferromagnetic and has a Curie point at 47°C. The magnetic moment of the manganese atom in this compound is M = 2.57 MV.

Vrzhetsiono [233] employed the powder-figure method to investigate the domain structure of Mn_5Ge_3. He found that the observed structure of this compound has uniaxial magnetic anisotropy; the axis of ready magnetization coincides with the hexagonal axis.

Alloys lying in the homogeneity region of the compound Mn_3Ge_2 have low hardness and are easily torn during rolling. Alloys having a γ-Mn structure are plastic and easily pressureworked.

Fakidov et al. [234, 235] investigated alloys containing from 25.2 to 80 at-% Ge. The alloys obtained proved to be ferromagnetic, but the Curie temperatures for all alloys containing from 25 to 33.3 at-% Ge were identical, lying in the vicinity of +25°C. The ferromagnetism of these alloys is due to the existence of a single ferromagnetic phase, apparently Mn_5Ge_2. The Curie

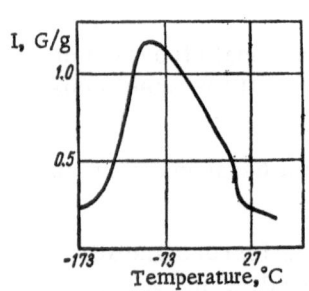

Fig. 35. Magnetic moment I of alloy Mn_3Ge_2 as a function of temperature [236] ($H = 1260$ Oe).

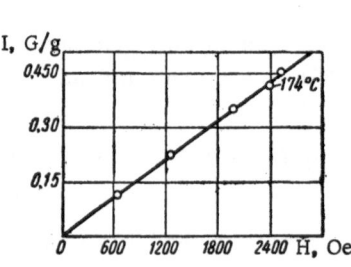

Fig. 36. Magnetic moment I of alloy Mn_3Ge as a function of magnetic-field intensity H [236].

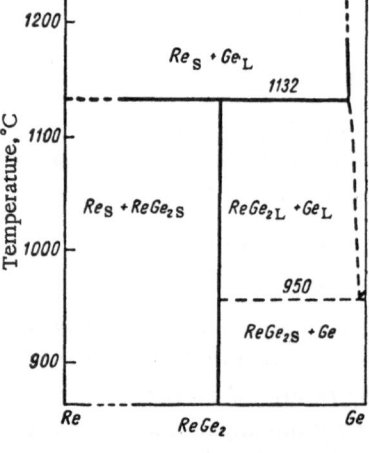

Fig. 37. Rough phase diagram of the rhenium—germanium system [222].

temperature for alloys containing 37-40 at-% Ge is + 50°C and Mn_5Ge_3 is the ferromagnetic carrier over this temperature range. Manganese—germanium alloys with a higher germanium content (more than 50 at-%) have two Curie points: -185 and -12°C.

Fakidov and Tsiovkin [236] made a further study of the magnetic properties of alloys with high germanium contents and determined the nature of the ferromagnetic transformations in Mn_3Ge_2. They established that Mn_3Ge_2 is a ferromagnetic with a Curie ferromagnetic point θ_f at 10°C and a paramagnetic point θ_p at 27°C. The magnetic moment of the manganese atom in this compound, calculated from paramagnetic measurements, is 2.5 MV. Figures 35 and 36 show the magnetic moment of the alloy Mn_3Ge_2 as a function of temperature and magnetic-field intensity.

The compound Mn_3Ge_2 exhibits a type I phase transition at -160°C [236].

Rhenium Germanides. Searcy et al. [222] employed x-ray analysis and vapor-tension measurements to establish the existence of the intermetallic compound $ReGe_2$ [43.79% Ge)

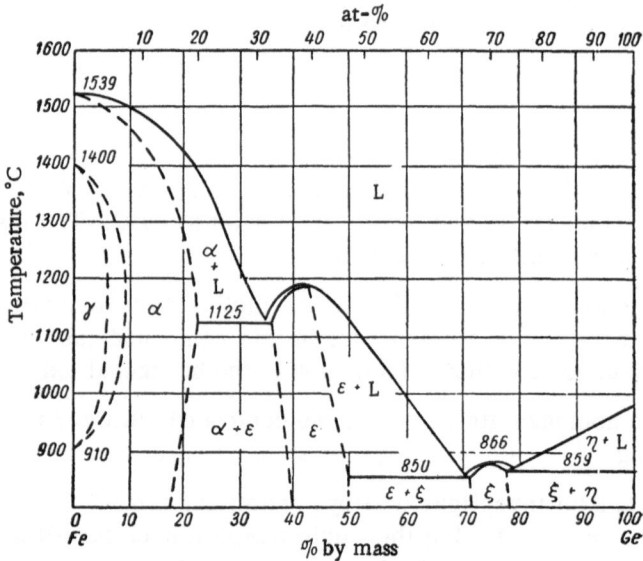

Fig. 38. Phase diagram of iron—germanium system [135].

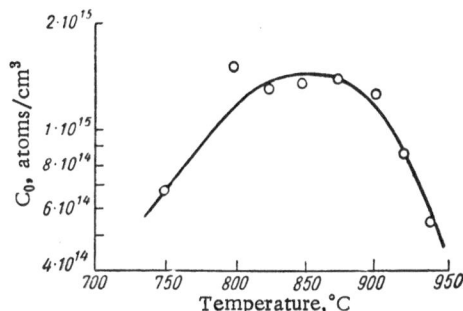

Fig. 39. Solubility of iron in germanium at 750 and 940°C [169].

which has a complex crystal structure. It decomposes peritectically at 1132°C to form solid rhenium and almost pure liquid germanium. Its heat of formation is $-\Delta H = 8.37$ kJ/mole (2 kcal/mole).

A rough phase diagram for this system is given in Fig. 37.

GERMANIDES OF IRON-GROUP METALS

Iron Germanides. Ruttewit and Masing [135], who investigated germanium−iron specimens prepared by fusion of germanium and iron in a nitrogen atmosphere, established the existence of γ-, α-, ε-, ξ-, and η-phase (Fig. 38).

The γ- and α-phase are solid solutions of germanium in iron. The ε-phase is a solid solution based on the germanide Fe_2Ge (39.39% Ge) and has a hexagonal structure of the NiAs type with the lattice constants a = 0.4039, c = 0.5032 nm, and c/a = 1.246 for an alloy with the composition $Fe_{1.76}Ge$ (42.47% Ge) [129] and a = 0.4017, c = 0.5005 nm, and c/a = 1.246 for an alloy with the composition $Fe_{1.7}Ge$ (43.33% Ge).

The ξ-phase is based on the germanide $FeGe_2$ (72.21% Ge) and has a tetragonal structure of the CuA_{12} type with the lattice constants a = 0.5899, c = 0.4941 nm, and c/a = 0.838 [238]. Crystallization of the ξ-phase is suppressed during rapid hardening of alloys containing more than 60 at-% Ge and a metastable $\varepsilon + \eta$ eutectic is formed at a Ge content of about 75% (70 at-%) and a temperature of 830°C.

The solubility of germanium in α-Fe is about 17% at 700°C.

Wever [239] confirmed the existence of a closed γ-region in this system.

Raub and Plate [223] confirmed that chemical compounds are present in the system. When specimens pressed from mixtures of powdered germanium and iron were heated in a vacuum at 1000°C, their linear dimensions changed. A chemical compound was formed at 862°C.

Barbier-Andrieux [88, 89], who studied the x-ray patterns of alloys containing 36.5% Ge prepared by electrolysis of molten sodium germanate, ferric oxide, and sodium fluoride, observed lines for a new γ'-phase with a cubic structure of the Cu_3Au type. According to his data, the solubility of germanium in α-Fe is 11.5%.

Bugai et al. [169] determined the solubility of iron in germanium over the temperature range 750-840°C. They found that the maximum solubility of iron in germanium at 850°C is $1.5 \cdot 10^{15}$ atoms/cm^3 (Fig. 39).

The germanium−iron alloys obtained by Barbier-Andrieux [88, 89] were porous and exhibited metallic fracture. The $FeGe_2$ alloy (ξ-phase) was nonmagnetic, while the Fe_2Ge alloy (ε-phase) and alloys richer in iron were highly ferromagnetic. The germanium−iron alloys underwent a magnetic transformation at 490°C.

The iron−germanium system was most completely studied by Shtol'ts et al. [121-125], who employed metallographic, x-ray, densitometric, thermomagnetic, and pyrometric analysis. They established that, in addition to the previously known phases (solid solutions of germanium in α- and γ-iron, the intermetallic compounds Fe_5Ge_3 [Fe_2Ge] and $FeGe_2$, and iron-alloyed germanium), there are four phases with the compositions $Fe_{3.25}Ge$, Fe_3Ge, Fe_4Ge_3, and $FeGe$.

TABLE 7. Certain Properties of Solid Solutions of Germanium in Iron (After Heat Treatment at 1000°C)

Germanium content of alloy, at-%	Lattice constant, nm	Experimental density, g/cm³	Microhardness		Curie point, °C
			MN/m²	kg/mm²	
2	0.2867	7.87	1200	120	747
4	0.2870	7.89	1400	140	744
6	0.2873	7.91	1750	175	740
8	0.2877	7.94	2150	215	735
10	0.2380	7.97	2800	280	725
12	0.2382	8.01	3200	320	712
14	0.2383	8.06	3700	370	695
16	0.2383	8.12	4350	435	677

On dissolving in α-Fe, germanium forms substitution solutions with a tendency toward ordering, which occurs by the mechanism of type-I phase transitions (as in the Cu_3Au alloy). The solubility of germanium is about 14 at-% at 1100°C and drops slightly, to about 12 at-%, when the temperature is reduced to 900°C. The constant of the cubic α-Fe lattice increases from 0.2866 to 0.2882 nm (in the alloy quenched from 900°C) or to 0.2883 nm (in the alloy quenched from 1100°C).

The physical properties of the α-solution are substantially altered when the germanium content is raised. Thus, the Curie temperature of alloys falls from 760°C for pure iron to 670°C for an alloy containing 16 at-% Ge (quenching from 900°C) or to 630°C (quenching from 1100°C). On the other hand, the resistivity, microhardness, and density increase with the germanium content (Table 7). Specimens containing 6 at-% Ge are easily rolled, but alloys with a higher germanium content become brittle.

Shtol'ts and Gel'd [124] give the results of thermomagnetic investigations of germanium—iron alloys. In their opinion, the α- and β-phases are in equilibrium in alloys containing from 85 to 64% Fe. As a result, the thermomagnetic curves exhibit two Curie points: $\theta_\alpha = 365°C$ and $\theta_\beta = 205°C$. Only one Curie point (205°C) is observed at an iron concentration of 64% and, when the iron content is further reduced, it drops almost linearly to 100°C at 55% Fe. On this basis, it is concluded that the β-phase is homogeneous when it contains between 55 and 64% Fe.

These results require refinement, since the authors did not make the necessary checks of system equilibrium. In consequence, they erroneously assumed that the β-phase can be in equilibrium with the α-saturated solid solution, not taking into account the existence of the intermediate germanide $Fe_{3.25}Ge$ [124, 125, 129].

The compound Fe_5Ge_3 (37.5 at-% Ge) has a hexagonal lattice of the NiAs type, whose constants vary little when the iron content of the phase is raised from 57.0 to 62.5% but undergo substantial changes when it is further increased. Within the homogeneity region (after quenching from 850°C), a increases from 0.3978 to 0.4034 nm and c from 0.4993 to 0.5024 nm. The compound Fe_5Ge_3 melts congruently [123, 124] at 1170°C and forms eutectics with $FeGe_2$ (at 35.5 at-% Fe and $t_e = 845°C$) and $Fe_{3.25}Ge$ (at 66 at-% Fe and $t_e = 1130°C$). This compound has a rather broad homogeneity region, which depends on temperature and is roughly characterized by the equations:

a) for the iron-saturated Fe_5Ge_3 phase: at-% $Fe = 63.4 + 0.0034t$ (from 500°C to $t_e = 1130°C$);

b) for the germanium-saturated Fe_5Ge_3 phase:

at-% $Fe = 58.3 - 0.003t$ (from 500°C to $t_e = 845°C$);

at-% $Fe = 57.6 + 0.016t$ (from 845 to 1170°C).

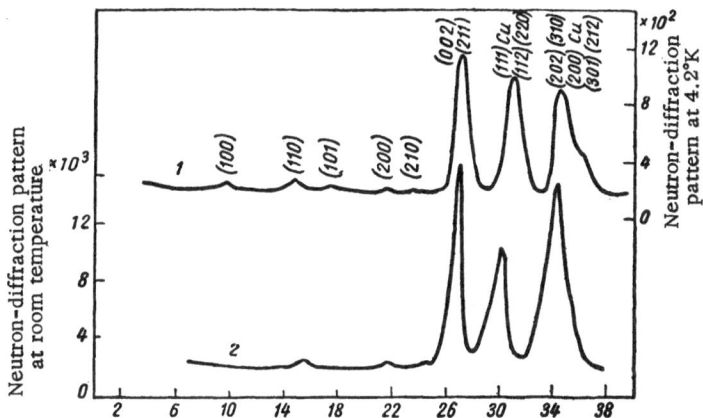

Fig. 40. Neutron-diffraction pattern of $FeGe_2$, obtained at 4.2°K [243, 244]. 1) 4.2°K 2) room temperature.

Investigation of the temperature function of resistivity for specimens in the homogeneity region of the Fe_5Ge_3 phase showed that this phase has a tendency toward ordering at temperatures below 850°C, retaining the same lattice. Alloys corresponding to the homogeneity region exhibit metallic conductivity and ferromagnetism. Their Curie temperature rises nonlinearly from 20 to 210°C.

At 750°C, the compound Fe_5Ge_3* (Fe_2Ge) passes from the brittle into the plastic state; this transition occurs within a narrow temperature range, 10-20°C. Above 750°C, Fe_5Ge_3 can undergo plastic deformation (bending and elongation).

The compound $FeGe_2$ (66.67% Ge) has a structure of the $CuAl_2$ type with the constants a = 0.5906, c = 0.4950 nm, and c/a = 0.838, which is in agreement with Wallbaum's data [238]. It melts congruently at 866°C and has a density of 7.7 g/cm^3 (i.e., 12 atoms, or four molecules, in the elementary cell) and a very narrow homogeneity region. The resistivity of this phase at room temperature is of the order of 10^{-4} ohm·cm and has a positive temperature coefficient ($\alpha = 10^{-3}$ 1/deg), which is halved as the temperature is raised from 20 to 400°C. Its thermal emf with respect to platinum is 0.4 μV/deg. According to Dudkin et al. [241, 242], $FeGe_2$ has metallic conductivity.

Forsyth et al. [243] and Kren and Szabo [244] employed neutron-diffraction analysis of polycrystalline and monocrystalline $FeGe_2$ specimens to determine the magnetic structure of this compound and investigate its hyperfine-interaction field. According to their data, the elementary antiferromagnetic $FeGe_2$ cell has the same dimensions as the elementary chemical cell, i.e., a = 0.5909, c = 0.4955 nm, and c/a = 0.838 [244].

Figures 40 and 41 show a neutron-diffraction pattern obtained at 4.2°K and a diagram of the elementary antiferromagnetic cell of the compound $FeGe_2$.

It was established that the spins of the four equivalent iron atoms in this cell are parallel or antiparallel in two different (but related by a symmetry operation) directions in the basal plane of the tetragonal structure. The angle between the two spin directions is 1.24 ± 0.1 rad (71 ± 6 deg), while the magnetic moment for each iron atom is 0.73 μV. The temperature function of magnetic neutron scattering indicates that the Newell temperature of the compound $FeGe_2$

---

*D. P. Shashkov, Author's abstract of dissertation [in Russian], Moscow (1965).

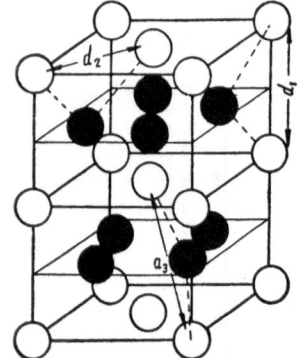

Fig. 41. Elementary anti-
ferromagnetic cell of the
compound $FeGe_2$ [243, 244].

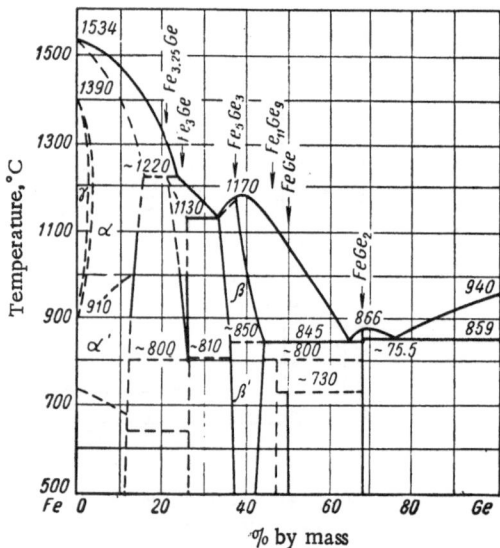

Fig. 42. Refined phase diagram of the
Fe–Ge system [245]

lies in the vicinity of 270°K. According to data obtained in measuring the Mossbauer effect the
hyperfine-interaction field at the poitions of the iron atoms is 148 ± 2 kOe.

The compound $Fe_{3.25}Ge$ ($Fe_{13}Ge_4$) is a solid solution with iron atoms replacing the germanium atoms in the Fe_3Ge elementary cell and is formed peritectically at about 1220°C. At 1000°C,
it is homogeneous at germanium concentrations between 23.3 and 24.5 at-%. It is unstable below 800°C and undergoes eutectoid decay into a Fe_3Ge phase and an ordered solid solution of
germanium in α-Fe. The compound $Fe_{3.25}Ge$ has a hexagonal lattice of the Mg_3Cd type (DO_{19}
space group) with constants that rise somewhat over the homogeneity region (at 1000°C), ranging from a = 0.5182, c = 0.4228 nm, and c/a = 0.8159 for an alloy containing 23.3 at-% Ge to
a = 0.5185, c = 0.4234 nm, and c/a = 0.816 for an alloy containing 25.0 at-% Ge; it has eight
atoms in the elementary cell and a density of 8.1 g/cm³.*

The physical properties of monophasic alloys in the homogeneity region are altered only
slightly by quenching from 1000°C. The resistivity of $Fe_{3.25}Ge$ is about $2-3\cdot10^{-4}$ ohm·cm. Its
microhardness is 5300 MN/m² (530 kg/mm²) and its thermal emf with respect to platinum is
4 µV/deg.

The compound Fe_3Ge has a face-centered cubic lattice of the Cu_3Au type with the constant
a = 0.3574 nm. It has a narrow homogeneity region and is ferromagnetic, with a Curie temperature of 530-550°C.

Metallographic and x-ray studies [124] of iron–germanium alloys containing from 45 to 65
at-% Ge showed that the phase compositions are materially altered by heat treatment under different conditions. The alloys were fused from monocrystalline germanium and iron with a purity of greater than 99.9% in sealed quartz ampules filled with argon to a pressure of about 13.33
N/m² (0.1 mm Hg). Before being analyzed, the specimens were homogenized and then heat-treated
in a vacuum of the order of 133.3 mN/m² (10^{-3} mm Hg). Alloys annealed at 800°C for 100 h consisted of two germanides, Fe_2Ge and $FeGe_2$. Specimens annealed at 600°C for 50 h underwent
a severe change in phase composition, alloys containing 49.51 at-% Ge being almost monophasic.
This investigation gives us grounds for surmising that the new low-temperature phase has a

---

*A. K. Shtol'ts, Author's abstract of dissertation [in Russian], Sverdlovsk (1963).

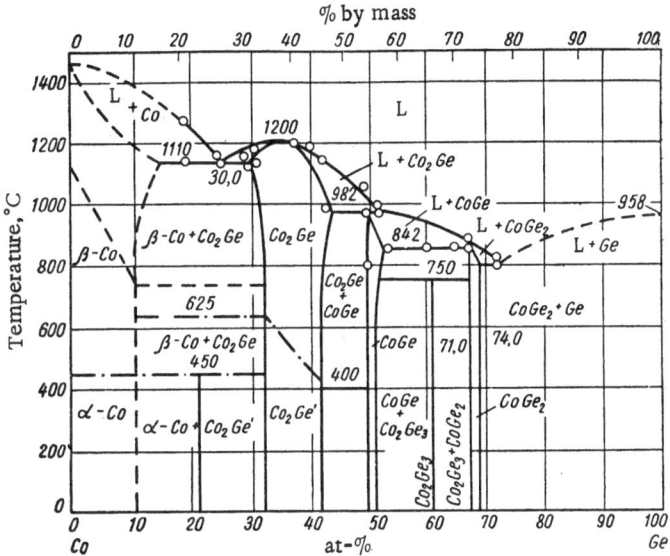

Fig. 43. Phase diagram of the cobalt–germanium
system [130].

composition corresponding to that of the monogermanide FeGe, which is formed by a peritectic reaction at a temperature somewhat below 800°C. According to Shtol'ts and Gel'd [124], the compound has a hexagonal lattice of the D_{6h}^1-C6/mmn space group. The germanium and iron atoms in the monogermanide lattice occupy the following positions: Ge atom 1 is in the 000 position, Ge atom 2 is in the 1/3, 2/2, 1/2; 2/3, 1/3, 1/2 position, and Ge atom 3 is in the 1/200, 01/20, 1/2, 1/20 position. The lattice constants of the FeGe elementary cell are a = 0.500, c = 0.4054 nm, and c/a = 0.810.

These data are in agreement with those of the Japanese researcher Ohoyama [127], who found FeGe in specimens prepared by heating the mixed components in sealed quartz ampules at 1300°C for 2 h, annealing them, and quenching them from a temperature below 700°C. He obtained the lattice constants a = 0.5003, c = 0.4055 nm, and c/a = 0.810. He believes that the compound is formed as a result of the peritectic reaction FeGe ⇌ β + δ at 700°C and has a narrow homogeneity region.

Study of the magnetic susceptibility of this compound as a function of temperature showed that the susceptibility curve has two temperature points, at 340 and 410°K. The microhardness of the FeGe phase is 8700 MN/m^2 (870 kg/mm^2).

Richardson [477] obtained single crystals of FeGe with a B20 structure by the iodide method, employing precipitation at 450°C in a quartz tube. The lattice constant was a = 0.4698 nm and the atomic parameters were x_{Fe} = 0.135 and x_{Ge} = 0.840. On heating to 740-756°C, the B20 structure was converted to a monoclinic (C2/m) structure with the constants a = 1.21, b = 0.393, c = 0.503 nm, and β = 103.6°C. This modification has a deficiency of at-% Ge in comparison with the equiatomic composition. Single crystals of the monoclinic modification were prepared by the bromide method, with precipitation in a quartz tube at 742°C, the reaction taking the form $FeBr_2 + 2GeBr_2 = FeGe + GeBr_4 + Br_2$.

The enthalpy and heat capacity of the solid and liquid germanides Fe_5Ge_3, Fe_4Ge_3, FeGe, and $FeGe_2$ were determined by Serebrennikov et al. [455] and by Shchipanova et al. [479].

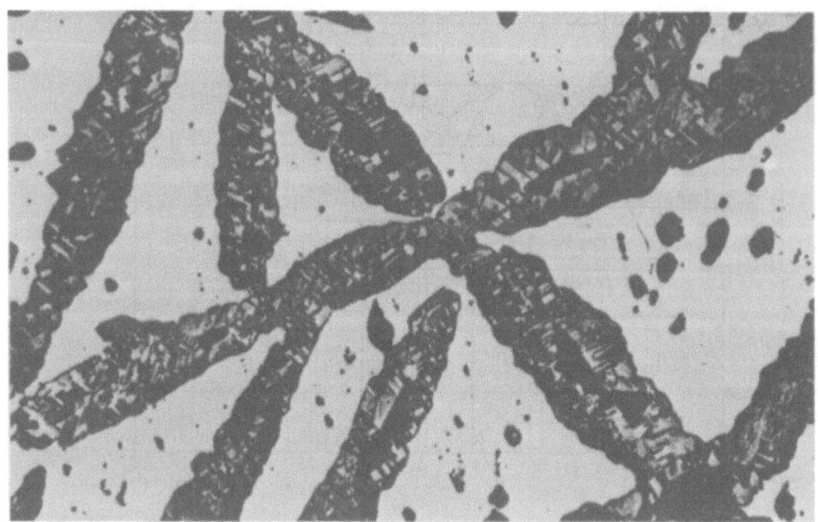

Fig. 44. Microstructure of Co−Ge alloy (Co_5Ge_7) [132].

The enthalpy of monophasic Fe_5Ge_3 increases monotonically with temperature over a rather broad range (0–1100°C) and is satisfactorily described by the equation

$$\Delta H_{273°}^T = -73{,}887 + 239.7T + 17.07 \cdot 10^{-3}T^2 + 1.972 \cdot 10^6 T^{-1},$$

whence it follows that the heat capacity (C_p) increases comparatively rapidly with temperature, conforming to the expression

$$C_p = 239.7 + 34.14 \cdot 10^{-3}T - 1.972 \cdot 10^6 T^{-2}.$$

The enthalpy of Fe_5Ge_3 varies linearly with temperature above its melting point:

$$\Delta H_{273°}^T = -35{,}710 + 333.4T;$$

whence $C_p = 333.4$ kJ/kmole·deg.

The heat of fusion ($\gamma_{Fe_5Ge_3}^f$), calculated by extrapolation of the temperature functions $\Delta H_{273°}^T$ for the solid and liquid phases and determination of the jump in enthalpy at 1146°C, was found to be 147,000 kJ/kmole.

The enthalpy of $FeGe_2$ also increases monotonically as the temperature is raised from 0 to 850°C and can be described by the equation

$$\Delta H_{273°}^T = -18{,}537.3 + 66.32T + 0.01552T^2 - 0.2012 \cdot 10^6 T^{-1},$$

whence

$$C_p = 66.32 + 0.03104T + 0.2012 \cdot 10^6 T^{-2}.$$

The enthalpy of $FeGe_2$ rises abruptly at 850°C, in conjunction with its melting. The fact that the melting range is very narrow despite the two-phase character of the specimen serves as additional confirmation of the small homogeneity region of $FeGe_2$ and the small slope of the

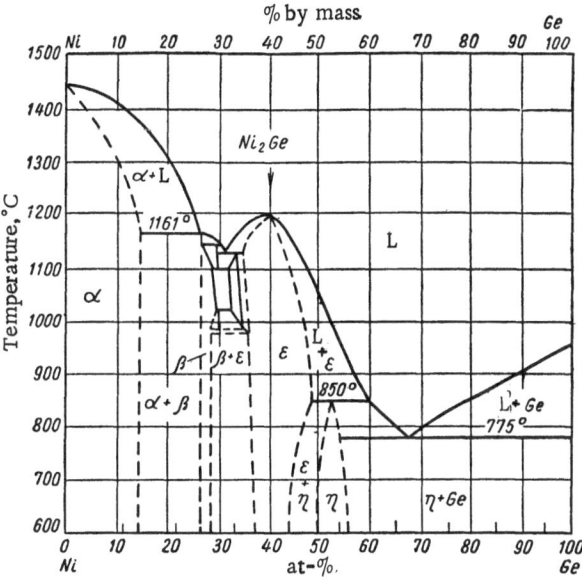

Fig. 45. Phase diagram of the nickel—germanium system [135].

solidus corresponding to it. The heat of fusion of FeGe$_2$ is γ_{ex}^{f} = 70,889 kJ/kmole. The enthalpy of liquid FeGe$_2$ rises linearly with temperature above the melting point:

$$\Delta H_{273°}^{T} = 23,930 + 109.5T.$$

Hence it follows that the heat capacity of liquid FeGe$_2$ is independent of temperature and equals 109.5 kJ/kmole·deg.

The variation in the enthalpies of Fe$_4$Ge$_3$ and FeGe was studied over the temperature range 0–1200°C.

The enthalpy of Fe$_4$Ge$_3$ increases monotonically as the temperature is raised to about 755°C and its temperature function is described by the equation

$$\Delta H_{273°}^{T} = -38,347 + 152.6T + 3.89 \cdot 10^{-2}T^2.$$

A peritectoid transformation takes place near 755°C, its thermal effect amounting to about 6100 kJ/kmole. A further rise in temperature to 845°C is again accompanied by a smooth increase in the enthalpy of the mixture of Fe$_5$Ge$_3$ and FeGe$_2$ crystals formed. The change in enthalpy under these conditions is represented by the equation

$$\Delta H_{273°}^{T} = -87,135 + 245T.$$

The enthalpy of the alloys increases isothermally to 9100 kJ/kmole at 845°C. This thermal effect corresponds to formation of a liquid eutectic, with eventual complete solution of the FeGe$_2$ crystals.

A further increase in system temperature leads to gradual solution of the Fe$_5$Ge$_3$ crystals and an increase in the amount of melt until 1080°C is reached, which is in satisfactory agreement with the phase diagram.

The system is monophasic above 1080°C and the enthalpy is described by the equation

$$\Delta H_{273°}^{T} = -7575 + 275T.$$

Extrapolation of the change in enthalpy as a function of temperature for the solid alloy to 1080°C enables us to estimate the heat of fusion of Fe_4Ge_3, which turns out to be 118,000 kJ/kmole.

The heat capacity of solid Fe_4Ge_3 between 0 and 755°C is described by the expression $C_p = 152.6 + 7.78T$, while that of liquid Fe_4Ge_3 is independent of temperature and equals 275 kJ/kmole·deg.

The temperature function of enthalpy for FeGe is described by the equations

$$\Delta H_{273°}^T = - 16,070.4 + 59.17T \text{ (over the range 0–735°C)},$$

$$\Delta H_{273°}^T = - 19,495 + 65T \text{ (over the range 755–845°C)}.$$

The effective heat of fusion of FeGe is 38,000 kJ/kmole. The change in enthalpy for the liquid phase is described by the expression

$$\Delta H_{273°}^T = - 9704 + 86.7T.$$

The heat capacities of solid and liquid FeGe are almost independent of temperature and equal 59.17 and 86.7 kJ/kmole·deg respectively.

On the basis of generalization of their data and the results of other investigations, Shtol'ts et al. [245] proposed a new version of the phase diagram of the iron−germanium system, which is shown in Fig. 42. This diagram requires careful checking and refinement.

Filatova et al. [246] investigated the influence of germanium on the hot strength of iron. For this purpose, alloys containing 3.75, 5.25, 7.5, and 20% Ge were prepared by vacuum induction fusion. The investigation was conducted by the differential manometric method described by Kubashevsky and Hopkins [247], using an air or oxygen atmosphere at a pressure of 36 kN/m^2 (270 mm Hg) over the temperature range 600–1000°C. This study established that the iron−germanium solid solution (with a Ge content of up to 20%) has a higher hot strength in the temperature region investigated than iron alone; the greatest effect was observed at temperatures of 700–800°C.

Cobalt Germanides. The phase diagram of the germanium−cobalt system was constructed on the basis of data obtained by thermal, microscopic, and x-ray analysis of alloys prepared by fusing 98% pure cobalt and germanium containing 0.001% impurities (Mn, Cu, and Cr) in a hydrogen atmosphere [130] (see Fig. 43).

The existence of the germanides Co_2Ge, $CoGe$, Co_2Ge_3, and $CoGe_2$ was established in this system; only Co_2Ge corresponds to an open maximum in the fusibility curve (at 1200°C). The germanide CoGe is formed at 982°C by a peritectic reaction between Co_2Ge and the melt, while the $CoGe_2$ is formed from the melt and CoGe at 842°C.

The compound Co_2Ge_3 is formed at 750°C by a peritectic reaction between CoGe and $CoGe_2$. It has almost no homogeneity region. All the other compounds have broad homogeneity regions and are capable of forming solid solutions.

In addition to the aforementioned germanides, Bhan and Schubert [131] discovered and studied the compounds CoGe and Co_5Ge_8.

The compound CoGe has a face−centered monoclinic structure of the Ni_3Sn_4 type (C_{2h}^3–C2/m space group) with the constants a = 1.1648, b = 0.3807, c = 0.4945 nm, and β = 1.77 rad (101.10°C) [131]. The elementary cell contains eight cobalt atoms and eight germanium atoms. The pycnometric density of the compound is 8.24 g/cm^3.

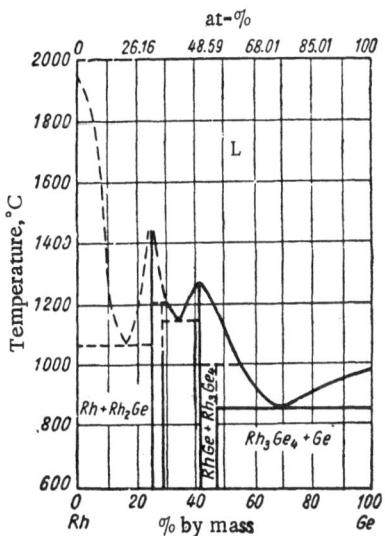

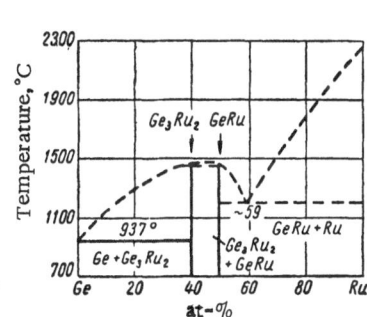

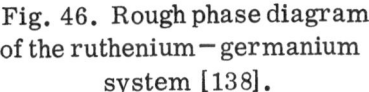

Fig. 46. Rough phase diagram
of the ruthenium—germanium
system [138].

Fig. 47. Phase diagram of the
rhodium—germanium system

The compound Co_5Ge_8 has a tetragonal structure with the constants a = 0.7641, c = 0.5814 nm, and c/a = 0.7608 [131].

In a more recent article, Stolz and Schubert [132] reported the existence of the compound Co_5Ge_7, whose lattice (C_{4V}^0-I4mm space group) has the constants a = 0.7641, c = 0.5814 nm, and c/a = 0.7608. Figure 44 shows the microstructure of the alloy.

Pfisterer and Schubert [130] established that the maximum solubility of germanium in β-Co is 17.85% (15 at-%) at 1110°C; the solubility of germanium in α-Co is 12.04% (10 at-%) and is independent of temperature.

Cobalt has not been found to be soluble in germanium. In alloys containing from 0 to 37 at-% Ge, α-Co is converted to β-Co at 450°C; the Co_2Ge phase is converted to its low-temperature modification, which is a superstructure (see Fig. 43), at 625 and 400°C. The coumpound Co_2Ge has a hexagonal structure of the NiAs type with the lattice constants a = 0.3917, c = 0.5045 nm, and c/a = 1.285. The lattice constants change when the germanium content is raised. Thus, the constants for alloys with a composition corresponding to the formula $Co_{1.76}Ge$ (41.16% Ge) are a = 0.3932, c = 0.5014 nm, and c/a = 1.275, while those for alloys corresponding to the formula $Co_{1.76}Ge$ (42.01% Ge) are a = 0.3917, c = 0.4979 nm, and c/a = 1.271 [136, 137].

The compound $CoGe_2$ is the first representative of a new type of crystal structure. At a Ge content of 72% (67 at-%), it has a face-centered rhombic, pseudotetrahedral structure of the $CoGe_2$ type with 16 germanium atoms and seven cobalt atoms in the elementary cell and the lattice constants a = b = 0.5681, c = 1.0811 nm, and c/a = 1.90 [170].

Raub and Plate [223], who measured the linear dimensions of a pressed micture (heated in a vacuum to 1000°C), established that an interaction between the metals to form a chemical compound begins at 825°C. The CoGe phase is brittle and its grain boundaries are not etched by nitric acid; its coloration (in polarized light) ranges from light yellow to dark violet.

Koester and Horn [248] established that the magnetic-transformation temperature of β-Co decreases to 720°C when 12% Ge is added. The density of an alloy containing 75.5% Ge is 7.50 g/cm^3. Table 8 presents some data on cobalt germanides.

Nickel Germanides. Ruttewit and Masing [135] detected the germanides Ni_3Ge, Ni_2Ge, and $NiGe$ in the germanium—nickel system (Fig. 45); only the Ni_2Ge phase corresponds to an open maximum in the fusibility curve at 1200°C. The solubility of germanium in solid nickel is about 13.8% at 1161°C. A rise in the dissolved-germanium content causes an increase in the constant of the cubic lattice of the α-solid solution. Thus, a rise from 1.20 to 6.65 at-% Ge increases the lattice constant from 0.3526 to 0.3535 nm, while a Ge content of 10.87 at-% yields a lattice constant of 0.3543 nm [249]. The solubility of nickel in solid germanium is extremely low. This was confirmed by Maesen and Brenkman [250]. The heat of solution of nickel in solid germanium is $\Delta H = 196.78$ kJ/mole (47 kcal/mole), while its temperature coefficient is $\Delta H/T = 71.17$ J/mole·deg (17 cal/mole·deg) [251].

Barbier-Andrieux [88, 89, 133], who studied alloys with Ni contents of up to 20.3% prepared by electrolysis of a melt containing sodium tetraborate, GeO_2, and NiO at 950-1000°C, observed only the phases NiGe and Ni_2Ge and a solid solution of germanium in nickel with a Ge content of 20.3-25.2%. The alloys were contaminated with nickel borides at an Ni content above 61.82%.

All the alloys had a metallic appearance. Those with Ni contents between 0 and 44 at-% were brittle. The alloy corresponding to the composition NiGe not only had a higher hardness than alloys richer in germanium, but was also less brittle. The alloy with the composition Ni_2Ge had an even higher hardness (scratching glass).

It has been established, however, that the compound Ni_3Ge passes from the brittle into the plastic state at 550°C. It can undergo plastic deformation (bending and elongation) above this temperature. According to Shashkov [485], this transition from the brittle to the plastic state in metallic compounds is due to disappearance of directional interatomic bonds.

Alloys lying in the region of solid solutions of germanium in nickel have magnetic properties [88, 89, 133].

Table 9 presents some data on nickel germanides.

PLATINOID GERMANIDES

Ruthenium Germanides. Wallbaum [82] reported the existence of the chemical compound $RuGe_2$ (58.94% Ge), which has a tetragonal structure of the $RuSi_2$ type.

Raub and Fritzsche [138], who investigated the germanium—ruthenium system by microscopic, x-ray, and thermal analysis, found that it contains the phases RuGe (41.65% Ge) and Ru_2Ge_3 (51.71% Ge).

The compound RuGe has a cubic structure of the FeSi type with the lattice constant a = 0.4846 nm.

The compound Ru_2Ge_3 has a tetragonal lattice of the Ru_2Si_3 type with constants a = 0.5709, c = 0.4650 nm, and c/a = 0.815. Neither compound has a noticeable homogeneity region. According to the aforementioned authors, germanium and ruthenium do not form solid solutions.

Figure 46 is a rough phase diagram of the germanium—ruthenium system.

Rhodium Germanides. Zhuravlev and Zhdanov [252] studied the rhodium—germanium system by microscopic and x-ray analysis and melting-point determinations. The phase diagram of this system is given in Fig. 47. Geller [83] had previously established the existence of the chemical compounds Rh_2Ge, Rh_5Ge_3, and RhGe. In addition to these compounds, Matthias [253] found the phase Rh_3Ge_2. Other investigations [252, 462] failed to confirm the existence of the chemical compound Rh_2Ge_2 in the system, instead demonstrating the presence of the phase Rh_3Ge_4.

Table 8. Properties of Cobalt Germanides

Germanide	Ge content, % by mass	Structure	Lattice constants, nm				Melting (decomposition) point, °C	Density, g/cm³
			a	b	c	c/a		
Co_2Ge	38.12	Hexagonal	0.3917	—	0.5045	1.285	1200	—
CoGe	55.19	Monoclinic	1.1648	0.3807	0.4045	$\beta = 1.77$ rad (101.10°)	982 (d)	—
Co_2Ge_3	64.88	—	—	—	—		759 (d)	—
$CoGe_2$	71.12	Face-centered rhombic	0.5681	0.5681	1.0811	1.90	842 (d)	7.50
Co_5Ge_8	66.33	Tetragonal	0.7641	—	0.5814	0.7608	—	—
Co_5Ge_7	68.29	''	0.7641	—	0.5814	0.7608	—	—

Table 9. Properties of Nickle Germanides

Germanide	Ge content, % by mass	Structure	Type	Lattice constants, nm				Melting (decomposition) point, °C
				a	b	c	c/a	
Ni_3Ge	26.8−29.0	Cubic	Cu_3Au	0.3567	—	—	—	1161 (d)
Ni_2Ge								
($Ni_{1.7}Ge$)	42.12	Hexagonal	NiAs	0.3919	—	0.5046	1.287 ⎱	1200
($Ni_{1.86}Ge$)	39.95	The same	NiAs	0.3954	—	0.5045	1.275 ⎰	
NiGe	55.29	Rhombic	MnP	0.5810	0.5380	0.2427	—	850 (d)

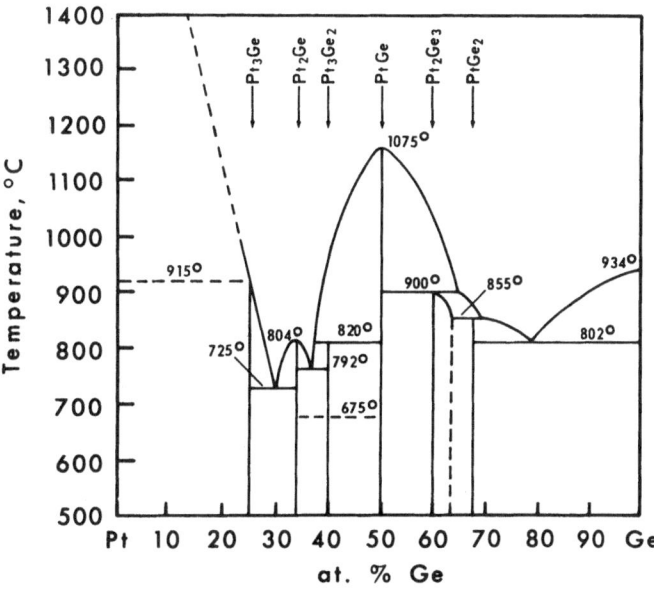

Fig. 47a. Phase diagram of the platinum−germanium system.

Table 10.   Properties of Rhodium Germanides

Germanide	Ge content, % by mass	Structure	Type	Space group	Lattice constants, nm				Melting point, °C	Density, g/cm³		Microhardness	
					a	b	c	c/a		Pycnometric	X-ray*	MN/m²	kg/mm²
Rh_2Ge	26.08	Rhombic	$PbCl_2$	Pnam	0.544	0.757	0.400	—	Over 1400	11.40	11.20	7000	700
Rh_5Ge_3	29.73	"	—	Pbam	0.542	1.032	0.396	—	—	10.60	11.00	4750	475
RhGe	41.36	"	MnP	Pbnm	0.570	0.648	0.325	—	1300	9.70	9.70	5800	580
Rh_3Ge_4	48.46	Tetragonal		—	0.570	—	1.00	1.754	—	8.50	—	5000	500

* The x-ray densities are from Geller [83].

Table 11. Structure and Certain Properties of Thorium Germanides

Germanide	Ge content, % by mass	Structure	Type	Space group	Lattice constants, nm				Melting point, °C	Density, g/cm³	
					a	b	c	c/a		Pycnometric	X-ray
Th_3Ge	9.44	—		—		—	—	—	—	11.00	—
Th_3Ge_2	17.26	Tetragonal	Th_3Si_2	D_{4h}^5	0.7951	—	0.4194	0.527	1800	10.48	10.54
ThGe	23.83	Cubic	NaCl	O_h^5	0.6046	—	—	—	1800	9.15	9.17
$ThGe_{1.5}$	31.93	Distorted rhombic	AlB_2	—	0.6989	0.8432	0.8136	—	1700	9.44	9.45
$ThGe_{1.6}$	33.35	The same	α-$ThSi_2$	—	0.5913	0.5889	1.4219	—	1650	—	9.34
$ThGe_2$	38.48	Rhombic	$ZrSi_2$	D_{2h}^{17}	0.4223	1.6911	0.4052	—	—	8.64	8.66
$Th_{0.9}Ge_2$	41.0	"	—	D_{2h}^{19}	1.6642	0.4023	0.4160	—	1600	—	8.44

Crystals of Rh_3Ge_4 were obtained by growth in a mother liquor and then isolated by dissolving the germanium. The resultant crystals were mostly concretions of irregular forms. The roentgenographically most perfect specimens were selected from a large number of such crystalline concretions. X-ray diffraction patterns established that the Rh_3Ge_4 crystals had tetragonal syngony and the vibration method was used to find their constants, which turned out to be a = 0.57, c = 1.00 nm, and c/a = 1.754.

The compounds Rh_2Ge and RhGe are formed with substantial evolution of heat and melt congruently, while the compounds Rh_5Ge_3 and Rh_3Ge_4 are formed as a result of peritectic reactions. According to Geller [83], the germanide Rh_5Ge_3 is a superconductor with a transition temperature of 2.12°K.

The solubility of one metal in the other is slight.

Table 10 presents some data on rhodium germanides.

Shchipanova et al. [479] established the existence of a new compound of electronic type $Rh_{17}Ge_{22}$. This compound has a tetragonal lattice (D_2^{12}-$\overline{1}42d$ space group) with the constants a = 0.560, c = 7.845 nm, and c/a = 13.998. The elementary cell contains 68 Rh atoms and 88 Ge atoms.

Palladium Germanides. Schubert el al. [134, 139, 254] established that, in addition to a solid solution of germanium in palladium, this system contains the chemical compounds Pd_4Ge (14.53% Ge), Pd_5Ge_2 (21.39% Ge), Pd_2Ge (25.39% Ge), and PdGe (40.49% Ge).

The compound Pd_4Ge is stable only at high temperatures.

The compound Pd_5Ge_2 is formed at low temperatures, as a result of ordering of the palladium-rich solid solution.

The compound Pd_2Ge has a hexagonal structure of the Fe_2P type with the lattice constants a = 0.6673, c = 0.3386 nm, and c/a = 0.507 [139, 254]. It melts congruently.

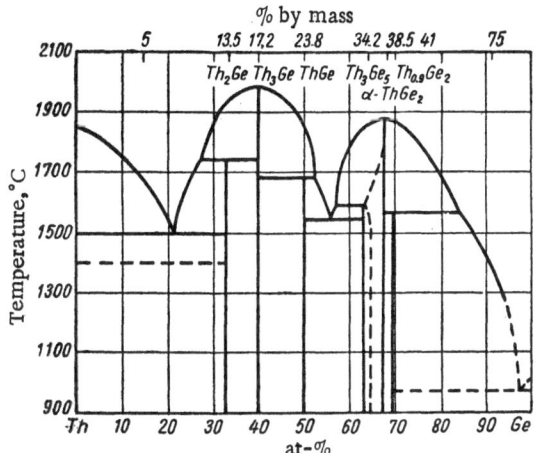

Fig. 48. Phase diagram of the Th—Ge system [149, 150].

The compound PdGe has a rhombic structure of the MnP type with the lattice constants a = 0.6258, b = 0.5781, and c = 0.3481 nm at a Ge content of 43% and forms a eutectic with the germanium at 70% Ge [134].

Schubert et al. [255] established the existence of the compound $Pd_{84}Ge_{16}$. This compound has a structure of the β-W type with the lattice constant a = 3157 nm.

Osmium Germanides. Osmium forms the germanide $OsGe_2$ (43.28% Ge), which is of the $RuSi_2$ type [82].

Weitz et al. [140] made a detailed study of the structure of this compound and found it to have the lattice constants a = 0.8995, b = 0.3094, c = 0.7685 nm, and α = 1.08 rad (119° 10'). Its pycnometric density is 11.10 g/cm^3, while its x-ray density is 11.90 g/cm^3.

Iridium Germanides. Pfisterer and Schubert [134], who studied alloys prepared by fusing germanium and high-purity (99.9%) iridium in evacuated quartz tubes, established the existence of the chemical compounds IrGe (27.41% Ge), which has a rhombic structure of the MnP type with the lattice constants a = 0.6280, b = 0.5611, and c = 0.3490 nm, and Ir_3Ge_7 (46.85% Ge), which has a body-centered cubic structure of the Ir_3Sn_7 type with 40 atoms in the elementary cell and the lattice constant a = 0.8752 nm.

The compound IrGe becomes superconductive at 4.7°K.

Bahn and Schubert [131] found two additional compounds, Ir_4Ge_5 and $IrGe_4$, in this system.

The compound Ir_4Ge_5 has a tetragonal structure with the lattice constants a = 0.564, c = 1.824 nm, and c/a = 3.234.

The compound $IrGe_4$ has a hexagonal structure with the lattice constants a = 0.6211, c = 0.7772 nm, and c/a = 1.251.

Platinum Germanides. According to microscopic and x-ray analysis of alloys prepared by fusion of these metals in evacuated quartz tubes, the system contains the compounds Pt_2Ge (15.68% Ge), PtGe (27.11 Ge), and Pt_2Ge_3 (35.81% Ge) [134].

The compound PtGe has a rhombic structure of the MnP type with the lattice constants a = 0.6088, b = 0.5732, and c = 0.3711 nm.

The compound Pt_2Ge has a hexagonal structure of the Fe_2P type. The lattice constants are a = 0.6683, c = 0.3527 nm, and c/a = 0.528 [139, 254].

Bahn and Schubert [131] found the chemical compounds $PtGe_2$, Pt_3Ge, and Pt_3Ge_2 and made a detailed study of their crystal structure.

The phase diagram of the Pt—Ge system (Fig. 47a) was constructed on the basis of microstructural, thermal, and x-ray analysis [494] of alloys fused from Pt (99.9%) and Ge (99.999%). The compounds Pt_2Ge and PtGe melt congruently at 800 and 1075°C respectively, while the compounds Pt_3Ge, Pt_3Ge_2, Pt_2Ge_3, and $PtGe_2$ are formed by peritectic reactions at 915, 820, 900, and 855°C respectively. The phase Pt_2Ge_3 is homogeneous over a narrow concentration region, the others having constant stoichiometric compositions. The compound Pt_3Ge_2 undergoes an allotropic transformation involving a strong thermal effect at about 675°C, the lattice constant b doubling (to 0.684 nm).

Formation of eutectics has been observed between Pt_3Ge and Pt_2Ge at 725°C and 30.5 at-% Ge, between Pt_2Ge and Pt_3Ge_2 at 792°C and 36 at-% Ge, and between $PtGe_2$ and Ge at 802°C and about 80 at-% Ge.

Table 1 gives the structures of platinoid germanides.

ACTINIDE GERMANIDES

Thorium Germanides. On the basis of an x-ray analysis of alloys prepared by sintering powders at temperatures between 1000 and 1500°C, Tharp et al. [148] reported the existence of six thorium germanides: $ThGe_{0.3 \pm 0.1}$, Th_3Ge_2, $ThGe$, $ThGe_{1.6 \pm 0.3}$ (β-$ThGe_2$), α-$ThGe_2$, and $ThGe_{3.0 \pm 0.4}$.

X-ray patterns of the compound $ThGe_{0.3 \pm 0.1}$ are quite complex.

The phase Th_3Ge_2 has a tetragonal lattice with the constants a = 7971, c = 0.4170 nm, and c/a = 0.523 (D_{4h}^5-P/mbm space group).

The compound ThGe has a cubic structure of the NaCl type with the lattice constant a = 0.6033 nm; it undergoes peritectic decay into Th_3Ge_2 and β-$ThGe_2$.

According to the data of Tharp et al. [148], the phase $ThGe_2$ exists in two modifications: α-$ThGe_2$ and β-$ThGe_2$. The α-$ThGe_2$ form is isostructural with α-$ThSi_2$ and α-USi_2, having a tetragonal structure with the lattice constants a = 0.4106, c = 1.4193 nm, and c/a = 3.456 (D_{4h}^{19}-14/amd space group); β-$ThGe_2$ has a distorted structure of AlB_2 type.

The $ThGe_{3.0 \pm 0.4}$ phase has been identified as having a pseudocubic structure with the constant a = 1.172 nm.

The thorium—germanium system has been most completely studied by Brown and Norreys [149, 150]. These authors employed metallographic and x-ray analysis and melting-point determinations to study the phase compositions and construct an experimental phase diagram (Fig. 48).

The basic method used to prepare the alloys was arc-fusion of the elements in a water-cooled copper crucible in an argon atmosphere, employing a tungsten electrode. In order to obtain a high degree of homogeneity, each alloy was refused several times.

Metallographic studies of the alloys showed that a thorium-rich phase (Th_3Ge) is formed as a result of the peritectoid reaction $2Th_3Ge \rightleftharpoons Th_3Ge_2 + 3Th$ at 1350-1400°C. A eutectic melting at 1450°C is formed between Th_3Ge_2 and Th at 12 at-% Ge. Alloys with a Ge content of 25 at-% consist of the single phase Th_3Ge.

The phase Th_3Ge_2 is formed at 40 at-% Ge and melts congruently at about 1800°C.

The compound ThGe was obtained at 50 at-% Ge. Alloys with a lower germanium content consist of a mixture of Th_3Ge_2 and ThGe. A eutectic mixture of these two phases is formed at 48-49 at-% Ge and has a melting point of about 1750°C.

The new phase $ThGe_{1.5}$ was found in alloys with a Ge content of about 60 at-%. The structure of this phase was not fully determined, but interpretation of the x-ray data gives some indication of the probable atomic positions. The Debye powder patterns of $ThGe_{1.5}$ are very similar to those for the defective C_{32} (AlB_2) structure of $ThSi_{1.67}$ (Th_3Si_5).

Specimens containing 60 and 61 at-% Ge exhibit a mixture of $ThGe_{1.5}$ and the phase $ThGe_{1.6}$, which has a structure of the C_c (α-$ThSi_2$) type. Alloys containing 61 and 63 at-% Ge produced by arc–fusion were monophasic and had a structure of the C_C type. However, heating of monophasic specimens with this composition at 900°C for 24 h led to precipitation of a second phase, $ThGe_{1.5}$ or $Th_{0.9}Ge_2$, depending on composition. Specimens containing 61.5 at-% Ge were monophasic after heat treatment and contained the phase $ThGe_{1.6}$.

The lattice constants of the solid solution rise from a = 0.5885, b = 0.5859, and c = 1.4219 nm for a monophasic alloy with a Ge content of 61 at-% to a = 0.5917, b = 0.5900, and c = 1.4225 nm for a diphasic alloy with a Ge content of 63–64 at-%.

Tharp et al. [148] designated the phase with a C_C structure as α–$ThGe_2$. The lattice constants reported for α–$ThGe_2$ are close to those obtained for $ThGe_{1.6}$. Although the values for α–$ThGe_2$ are lower than those for $ThGe_{1.6}$, Brown and Norreys [150] believe them to be the same phase.

Alloys containing 63–68 at-% Ge consist of a mixture of $ThGe_{1.6}$ with a phase containing 68–69 at-% Ge. The x-ray data for this phase indicate that it corresponds to the compound $ThGe_{3.0 \pm 0.4}$ described by Tharp et al. [148]. A detailed study has shown that this phase has a rhombic structure of the C_{49} type and the composition $ThGe_2$.

Alloys with a Ge content of more than 68 at-% are diphasic and consist of $Th_{0.9}Ge_2$ and Th. A eutectic mixture melting at 900°C has been obtained at a Ge content of 90 at-%.

Table 11 presents the data obtained by Brown and Norreys [150] on the structure and properties of thorium germanides.

Uranium Germanides. Iandelli and Ferro [197] obtained the uranium germanide UGe_3 (47.78% Ge), which has a structure of the Cu_3Au type and the lattice constant a = 0.4198 nm. Frost and Maskrey [256] found the lattice constant of UGe_3 to be 0.4206 nm. The melting point of this compound is about 1200°C and its x-ray density is 10.37 g/cm³. The data obtained by Frost and Maskrey [256] for the structure and lattice constant of UGe_3 were confirmed by Makarov and Bykov [257].

Lyashenko and Bykov [84] conducted a detailed investigation of the germanium—uranium system, employing thermal and microscopic analysis. Figure 49 shows the phase diagram of this system.

Microstructural analysis revealed the germanides U_5Ge_3 (37.5 at-% Ge), U_3Ge_4 (57.14 at-% Ge), UGe_2 (66.6 at-% Ge), and UGe_3 (75.0 at-% Ge). Moreover, the microstructure of an alloy containing 12.5 at-% Ge, which was quenched from 1000°C with prequenching holding times of 6, 126, 146, and 366 h, showed that the high-melting U_5Ge_3 grains dissolve in a eutectic with 3.5 at-% Ge and then form a chemical compound, which has been assigned the provisional formula U_7Ge_3. This phenomenon can readily be seen in the photomicrographs in Fig. 50.

The compound U_3Ge_4 has a rhombic structure of the Mn_5Si_3 type with the lattice constants a = 0.8577, c = 0.5791 nm, and c/a = 0.675. Its density is 13.40 g/cm³.

The compound U_3Ge_4 has a rhombic structure with the constants a = 0.5871, b = 0.9879, and c = 0.8978 nm.

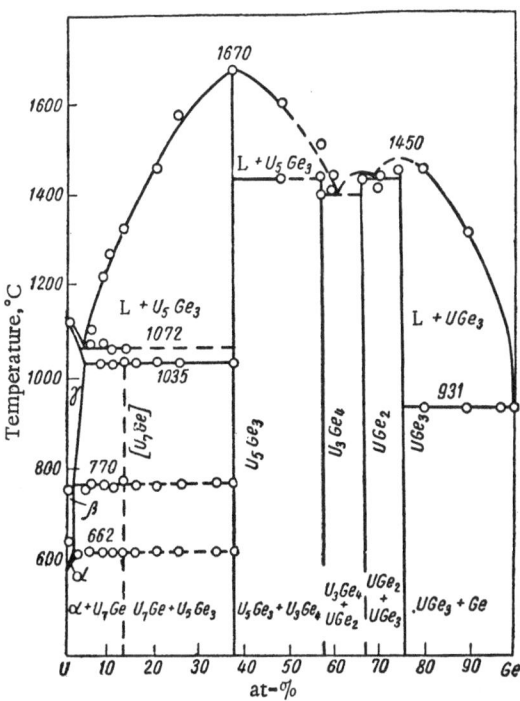

Fig. 49. Phase diagram of uranium—germanium system [84].

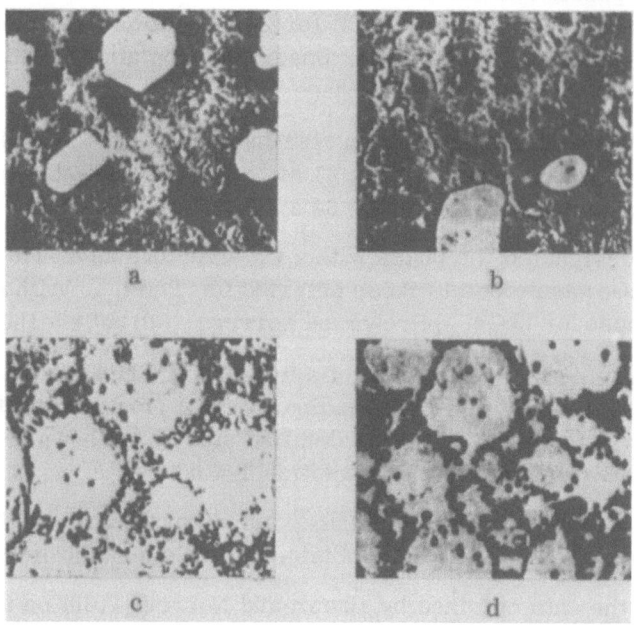

Fig. 50. Microstructure of uranium−germanium alloy containing 12.5 at-% Ge[84]. a) Quenching from 1000°C, holding time 6 h; b) quenching from same temperature, holding time 126 h; c) quenching from same temperature, holding time 366 h; d) quenching from same temperature, holding time 366 h (sintered alloy).

The compound UGe_2 has a rhombic structure of the $ZrSi_2$ type with the lattice constants a = 0.4118, b = 1.5130, and c = 0.3978 nm [257].

Study of the variation in lattice constant and microstructure in quenched alloys made it possible to establish that the maximum solubility of germanium in α- and β-uranium is about 1%, while that in γ-uranium ranges up to 3.0%. Uranium is almost insoluble in germanium.

When a pressed mixture of powdered germanium and uranium is heated in vacuum, a vigorous interaction between the metals begins at 310°C [223].

Plutonium Germanides. The existence of the germanides $PuGe_3$, $PuGe_2$, and Pu_2Ge_3 has been established in the germanium−plutonium system [258, 259]. The compound $PuGe_3$ has a cubic structure of the Cu_3Au type with the lattice constant a = 0.4223 nm. Its x-ray density is 10.07 g/cm^3.

The germanide $PuGe_2$ has a tetragonal structure of the $ThSi_2$ type with the lattice constants a = 0.4102, c = 1.381 nm, and c/a = 3.360 [259].

The compound Pu_2Ge_3 has a hexagonal structure (distorted AlB_2 type) with the lattice constants a = 0.3975, c = 0.4198 nm, and c/a = 1.056 [259]. The x-ray densities of these compounds are 10.98 and 10.06 g/cm^3 respectively.

CHAPTER 5

COMPOUNDS OF GERMANIUM WITH NONMETALS AND SEMIMETALS

COMPOUNDS WITH BORON-SUBGROUP ELEMENTS

Germanium – Boron System. The Ge–B system has been found to contain the compounds GeB_4 (37.35% B) [260] and GeB_6 [261].

The literature contains no data on the solubility of boron in solid germanium. Boron is an acceptor in germanium and can, in a number of cases, be considered as a contaminating impurity; however, it is sometimes used as a special alloying additive to give germanium hole-type conductivity.

The compound GeB_6 was obtained by vacuum-thermal reduction of germanium dioxide (GeO_2) with a mixture of boron carbide and carbon at 1200°C, by the reaction

$$2GeO_2 + 3B_4C + C = 2GeB_6 + 4CO.$$

The GeB_6 yield decreases as the temperature is raised and the entire reaction mass volatilizes at 1400°C.

Boron-carbide reduction in a porcelain reaction tube at 1200°C yields similar results.

The same products were contained in an attempt to produce germanium boride by reduction of germanium dioxide with boron in a vacuum at 1250°C for 30 min:

$$GeO_2 + 8B = GeB_6 + B_2O_2.$$

Attempts to obtain GeB_6 by synthesis from germanium and boron powders at temperatures of 800, 900, and 1200°C with holding times of 2.4, 8, and 1.5 h were unsuccessful.

An important characteristic of boron is the fact that it can be concentrated in the first portions of solid material formed during crystallization of molten germanium containing an admixture of this element.

According to Krasyuk and Gribov [2], the diffusion constant of boron in germanium is $D_0 = 1.8 \cdot 10^{-9}$ cm^2/sec and the activation energy of diffusion is $Q = 439.61$ kJ/mole (105 kcal/mole).

Aluminum – Germanium System. The phase diagram of the Ge–Al system, which was worked out by Stoehr and Klemm [263], is of the eutectic type and has a eutectic point at 55% Ge and 423°C. No germanides were detected in it (Fig. 51). These authors employed x-ray diffraction analysis of specimens annealed for several months to investigate the solubility of aluminum in germanium and of germanium in aluminum. They established that the x-ray patterns of specimens containing no less than 5 at-% Al exhibited weak aluminum interference lines at 395°C. It was found that the maximum solubility of aluminum in germanium is about

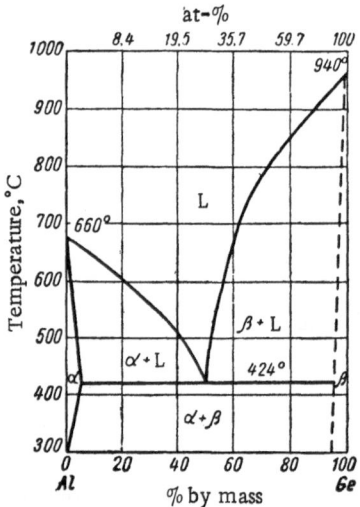

Fig. 51. Phase diagram of the aluminum−germanium sysem [263].

2−4 at-%, while that of germanium in aluminum is about 2.8 at-%, at 424°C. Maximum solubility decreases as the temperature is reduced and amounts to about 0.2 at-% Ge at 177°C.

Axon and Hume-Rothery [265] reported on the lattice constants of solid solutions containing up to 1.98 at-% (5.15% by mass) Ge. Germanium forms a solid solution with aluminum and increases the lattice constant of the latter by 0.00013 nm for each at-% Ge. Moreover, the data obtained show that at least 2 at-% Ge dissolves in aluminum at 500°C. Thurmond [163] demonstrated that the solubility of germanium in molten aluminum can be expressed by the rules for regular solutions. Germanium fuses very readily with aluminum, even at low temperatures. Aluminum−germanium alloys, like silumins, can be modified with sodium.

Boom [266], who investigated the modifying effect of sodium on alloys prepared from high-purity aluminum (99.99%) and germanium, concluded that small amounts of sodium create alloys with a finer-grained structure. The eutectic point is displaced toward a higher germanium content, while the crystallization temperature of the eutectic is reduced by about 8°C.

Sodium modification of an alloy containing 65% Ge causes a new phase (a ternary aluminum−germanium−sodium compound) to appear, in the form of light gray acicular crystals.

In an investigation of the modification of aluminum−germanium alloys by lithium, the same author [267] found that large light gray crystals of a ternary aluminum−germanium− lithium compound appear in the alloys.

Addition of from 0.041 to 0.39% germanium improves the tensile properties of rolled aluminum at ordinary and elevated temperatures (up to 425°C). The hardening at elevated temperatures is due both to formation of a solid solution and to a rise in the temperature at which the cold-hardened alloy begins to recrystallize [268].

The resistivity of alloys increases with rising germanium content, from 2.747 μohm·cm for pure aluminum to 2.860 μohm·cm for an alloy containing 0.39% Ge [269].

Gallium − Germanium System. Many researchers [181, 183, 270-275] have studied the gallium−germanium (Ga-Ge) system. Its phase diagram is eutectic in character, with a eutectic melting at 29.8°C (Fig. 52). The melting point of germanium and the solubility of gallium in germanium have been determined. Some authors report the melting point of germanium to be 958°C [270], while others hold it to be 237.2°C [271, 272]. The solubility of gallium in germanium was determined by x-ray diffraction analysis.

Greiner and Breidt [271, 272] found that solution of gallium in germanium leads to a rise in the lattice constant of the germanium [the increase is 0.00004 nm (0.0004 Å) per at-% Ga] and that the solubility of gallium is about 1 at-% at 600-900°C, 2.0 at-% at 600°C, and 2.5 at-% at 780°C.

Sheka et al. [276] determined the solubility of germanium in gallium at 200-500°C by suspending accurately weighed germanium specimens in molten gallium until a saturated solution was obtained. The amount of dissolved material was found from the change in melt weight; spectral analysis was employed for solubility levels of less than 0.1 at-%. Solution of germanium in

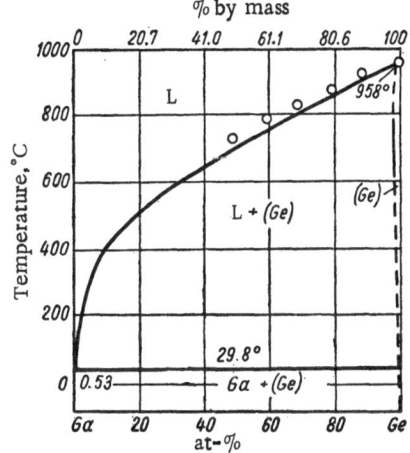

Fig. 52. Phase diagram of the gallium—germanium system [270].

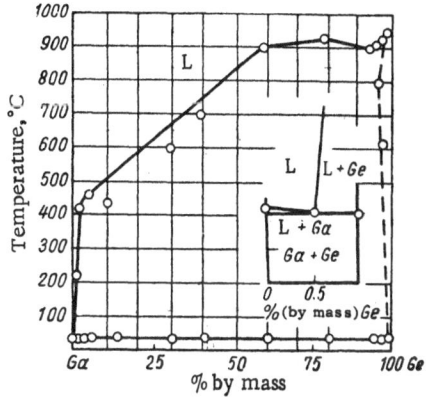

Fig. 53. Phase diagram of the gallium—germanium system [181,183].

gallium entails absorption of 34.75 kJ/mole (8.3 kcal/mole) [275]. The diffusion constants for gallium and germanium (at 800°C) are $D_0 = 1 \cdot 10^{-13}$ cm^2/sec and $D_0 = 20$ cm^2/sec respectively [153, 175]. The activation energy of diffusion is Q = 297.27 kJ/mole (71 kcal/mole).

The most complete investigation of the germanium—gallium system was conducted by Savitskii et al. [181, 183]. They employed the thermal and metallographic methods, as well as measurements of microhardness and electrical resistance at room temperature. The alloys were prepared by fusion of high-purity materials (99.9% Ga and 99.99% Ge) in evacuated and sealed quartz ampules. The results obtained were used to construct a phase diagram (Fig. 53), which differs from the diagram given in Fig. 52 in the more abrupt rise in the liquidus for gallium-rich alloys with increasing germanium content and the less abrupt drop in the liquidus for germanium-rich alloys with increasing gallium content. The diagram in Fig. 53 is more nearly correct, since it is based on the results of studies of many alloys by different physico-chemical analytic methods.

Figure 54 shows the variation in the resistivity of these alloys as a function of composition.

The resistivities of the initial components are $60 \cdot 10^{-6}$ ohm · cm for gallium and 6.8 ohm · cm for germanium. The microhardnesses of gallium, germanium, and the eutectic are 100–110 MN/m^2 (10–11 kg/mm^2), 6000–7000 MN/m^2 (600–700 kg/mm^2), and 355.9 MN/m^2 (35.59 kg/mm^2) respectively.

Thurmond et al. [67] determined the liquidus curves for the Ge—Ga, Ge—In, and other systems from the content of the second component (Ga, In, etc.) in germanium crystals deposited during slow cooling of molten alloys. The results of these experiments are shown in Fig. 55.

Addition of even small amounts of gallium reduces the hardness and resistance of germanium. Its resistance falls to $2.1 \cdot 10^{-3}$ ohm·cm at a Ga content of 2.5%. All the alloys in this system exhibit metallic conductivity. Alloys with semiconductive properties were not detected even in the germanium-rich region.

Sheka et al. [276] have also considered the Ga—Ge system.

Indium — Germanium System. Indium is widely employed as an acceptor alloying element to create individual n-type regions with hole conductivity in germanium crystals. Because of its low melting point, indium is preferable to such common acceptors as aluminum and is now used in the production of many germanium-based electronic devices.

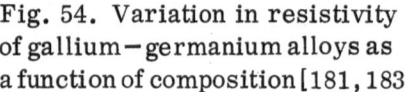

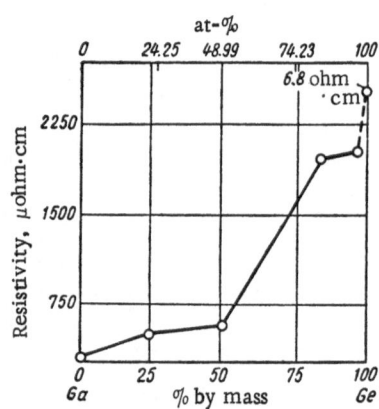

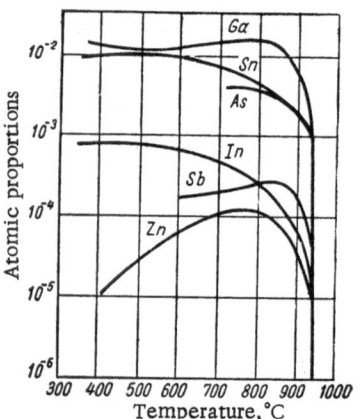

Fig. 54. Variation in resistivity
of gallium—germanium alloys as
a function of composition [181, 183].

Fig. 55. Solidus curves for
germanium-rich alloys [67].

A phase diagram for alloys in the Ge—In system has been constructed on the basis of data
obtained in thermal, metallographic, and x-ray diffraction analyses (Fig. 56) [270, 275]. The
liquidus and solidus curves and the composition of the eutectic in this system were determined
by Thurmond et al. [67, 163]. According to their data, the melting point of the eutectic lies 0.4°C
below that of indium and the germanium concentration in the eutectic alloy is 0.1 at-%. Figure
55 shows the solidus in the region of germanium-rich alloys. Germanium and indium are mutually soluble in all proportions in the liquid state (at a sufficiently high temperature).

Roschen and Thornton [277] established the boundaries of the region in which solid solutions of germanium in indium exist and found the eutectic point. The eutectic concentrations
proved to be 0.023 at-% Ge and 99.977 at-% In. The melting point of the eutectic lay very close
to that of pure indium. The solubility of indium in solid germanium at 20°C was $3 \cdot 10^{19}$ atoms/cm^3
[154]. These data are not in agreement with those of Hall [251], who conducted the following
experiment. A weighed portion of indium (20 g) was melted in a quartz tube in a hydrogen atmosphere. The resultant melt was saturated with germanium at temperatures between 400 and 600°C
and then cooled at a rate of 25 deg/h. The germanium crystallized out of the melt as the temperature was reduced. When the germanium crystals floating on the surface of the melt reached
a length of 2-3 mm, they were removed from the quartz tube. The surface indium was eliminated by
etching the crystals with hydrochloric acid. Etched crystals with a length of about 1 mm and a
diameter of 0.2 mm served as specimens for resistivity measurements. The resistivities found
were used to calculate the content of indium (which is a simple acceptor) in the germanium, employing the following relationship between the free current-carrier concentration at room temperature and the resistivity of germanium alloyed with an acceptor impurity [2]:

Charge-carrier concentration, cm^{-3}	$1 \cdot 10^{17}$	$1 \cdot 10^{18}$	$1 \cdot 10^{19}$
Resistance of Ge with hole conductivity, ohm·cm	0.057	0.011	0.0028

The average indium concentration in the germanium for crystals grown at 550°C was $2.6 \cdot 10^{18}$
atoms/cm^3, while that for crystals grown at 360°C was half this figure.

The heat of solution (-ΔH) of germanium in indium is 49.82 kJ/mole (11.9 kcal/mole), while
that of indium in germanium is 83.74 kJ/mole (20 kcal/mole) [275]. The temperature coefficient
of the heat of solution of indium in solid germanium (ΔH/T) is 30.56 kJ/mole·deg (7.3 cal/mole·
deg [251]).

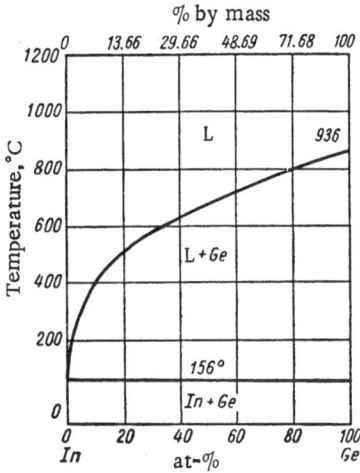

Fig. 56. Phase diagram of the indium—germanium system [270, 275].

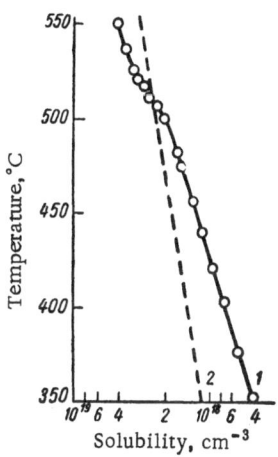

Fig. 57. Solubility of indium in solid germanium. 1) Khukhryanskii's experimental data [278]; 2) Trumbore's data [280].

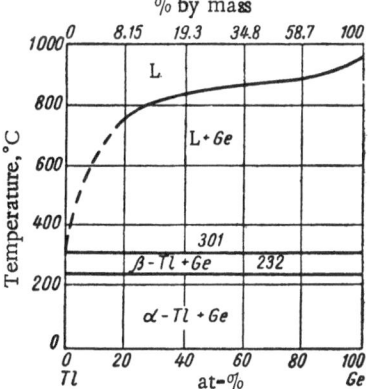

Fig. 58. Phase diagram of the thallium—germanium system [270].

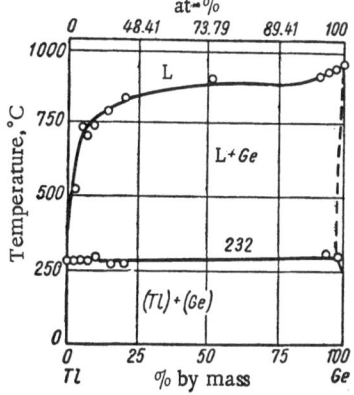

Fig. 59. Phase diagram of the thallium—germanium system (Savitskii [183]).

The value found by Hall for the solubility of indium in germanium is somewhat too high, since the crystals isolated contained indium inclusions, which caused a decrease in their electrical resistance. The solubility of indium in germanium has not as yet been conclusively determined.

Khukhryanskii [278] attempted to establish the solubility of indium in germanium over the temperature range 350–550°C by the method described in another of his articles [279]. Figure 57 shows the results obtained in his experiments (curve 1) and the data of Trumbore [280] (curve 2). Comparison of these data indicates that the solubility function for solid germanium had the same character in both cases, but Khukhryanskii's results showed the solubility of indium to depend to a greater extent on temperature, which is in better agreement with the theoretical curve for the solubility of indium in solid germanium [281].

Thallium — Germanium System. Klemm et al. [270] showed that the phase diagram of the Ge—Tl system is of the eutectic type. Its eutectic point lies very close to pure

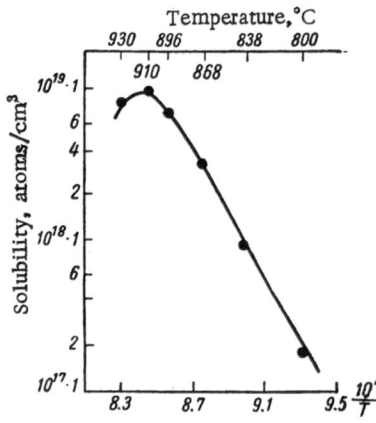

Fig. 60. Solubility of gallium in germanium as a function of temperature [282].

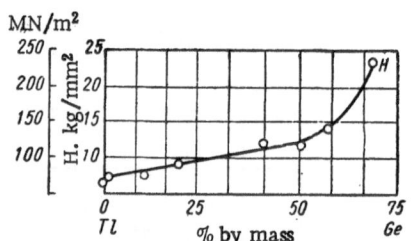

Fig. 61. Hardness of Ge−Tl alloys as a function of composition[183].

thallium and the melting point of the eutectic differs from that of thallium by only 1–2°C (Fig. 58). The form of the liquidus for this system indicates nonmiscibility in the liquid state, a considerable portion of the curve being hypothetically represented by a dash line.

Using the same methods they employed for the germanium−gallium system, Savitskii et al. [183] found that the primary-crystallization branch for germanium extends from the melting point of pure germanium to that of thallium. There is no primary crystallization branch for thallium. Addition of 85% Tl to germanium slightly reduces the melting point of the alloys (to 800°C). The eutectic point is displaced toward the melting point of thallium, so that this diagram can be considered to be of the pseudoeutectic type (Fig. 59).

Accoding to the data of Thurmond [163], the melting point of the eutectic in the germanium-thallium system lies 0.6°C below that of thallium, while the germanium concentration in the eutectic is 0.1 at-%.

Tagirov and Kuliev [282], who investigated the diffusion and solubility of thallium in germanium, found that the solubility of thallium reaches a maximum of $9.5 \cdot 10^{18}$ atoms/cm^3 at 917°C and then decreases, which is associated with its retrograde character (Fig. 60).

The hardness of thallium-rich alloys increases with their germanium content, from 30 MN/m^2 (3 kg/mm^2) for pure thallium to 90 MN/m^2 (9 kg/mm^2) at 50% Ge and 250 MN/m^2 (25 kg/mm^2) at 70% Ge (Fig. 61). The microhardnesses of the structural constituents are 60–70 MN/m^2 (6–7 kg/mm^2) for thallium and 6600 MN/m^2 (660 kg/mm^2) for germanium [183].

All the alloys in the system exhibit metallic conductivity. Figure 62 shows the variation in their electrical resistance as a function of composition.

COMPOUNDS WITH CARBON–SUBGROUP ELEMENTS

Germanium − Carbon System. Scace and Slack [283] studied the solubility of carbon in germanium. Zone-refined germanium was placed in a graphite container and heated to 2780–3170°C in the furnace shown in Fig. 63. After the furnace cooled, the germanium ingot was removed from the container and etched in a mixture of HF and HN0$_3$; the precipitate obtained, which had the form of fine flakes of graphite and some germanium, was analyzed for carbon content. The results of this experiment are shown in Fig. 64.

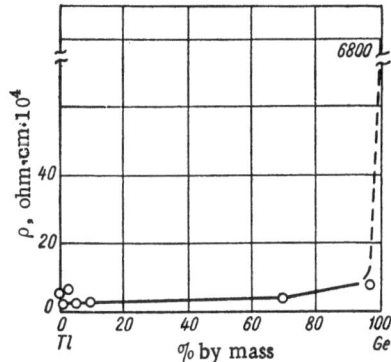

Fig. 62. Variation in electrical resistance of thallium−germanium alloys as a function of composition [183].

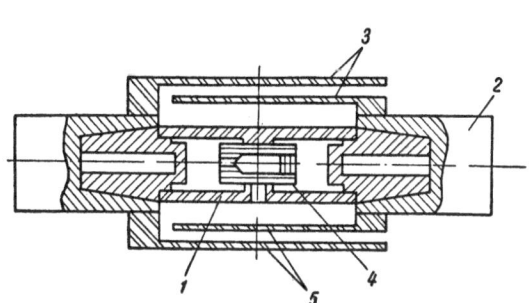

Fig. 63. Graphite resistance furnace [283]. 1) Heating element; 2) leads; 3) shields; 4) container; 5) inspection windows;

When the solubility curve was extrapolated to the melting point of germanium, the carbon content of the liquid germanium was found to be 10^8 atoms/cm^3. Formation of a compound between carbon and germanium was not detected until a temperature of 3170°C was reached. The phase diagram of the C−Ge system is of the simple eutectic type and resembles the diagrams for the C−Pb and C−Sn systems; the eutectic point lies close to the melting point of pure germanium (Fig. 65).

Silicon − Germanium System. The first serious investigation of the Ge−Si system was conducted by Stoehr and Klemm [263], who showed that a continuous series of solid solutions is formed between germanium and silicon [Fig. 66]. The same type of diagram was also reported by Barbier-Andrieux [284], who studied the structure of alloys containing from 0 to 48% Si prepared by electrolysis of molten lithium silicate and germanium dioxide at a temperature of 1250-1270°C.

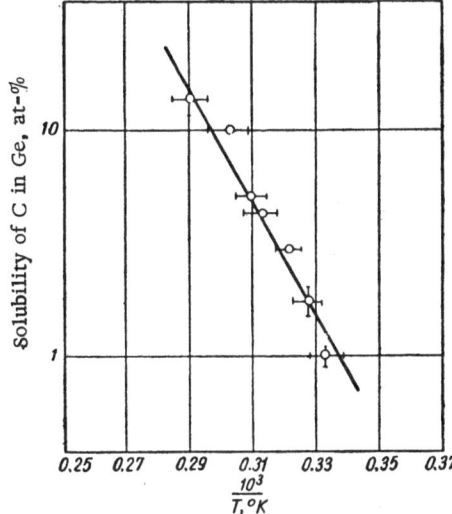

Fig. 64. Solubility of carbon in germanium as a function of temperature [283].

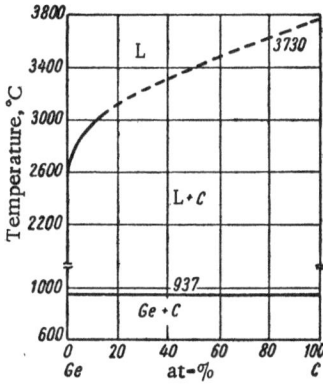

Fig. 65. Phase diagram of the carbon−germanium system [283].

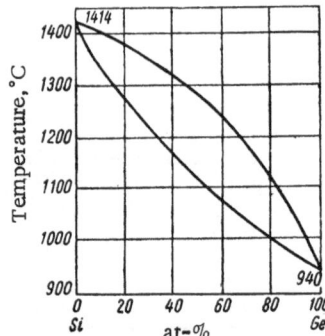

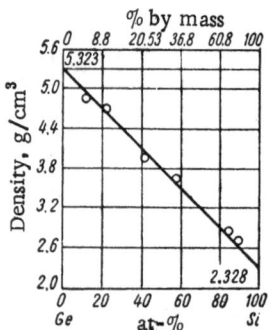

Fig. 66. Phase diagram of the silicon—germanium system [263].

Fig. 67. Variation in density of Ge—Si alloys as a function of composition [286].

A thermodynamic investigation of the system [163] confirmed the position of the solidus in the diagram in Fig. 66; it was found, however, that the liquidus should be somewhat lower than in this diagram. On the other hand, experimental determinations of the liquidus and solidus for alloys containing from 7.4 to 33.5 at-% Si [285] yielded results in good agreement with this diagram. Thus, the liquidus temperatures for alloys containing 7.4 and 33.5 at-% Si were found to be 1027 and 1207°C respectively, while the solidus temperatures for alloys containing 10, 13, and 16 at-% Si were found to be 963, 971, and 987°C respectively. Stoehr and Klemm [263] showed that, even though both components crystallize with the same diamond-type lattice and exhibit only a small difference (4%) in lattice constants, equilibrium is reached very slowly in the system, at a temperature 50 deg below the melting point of germanium. They consider the slight tendency of the diamond structure toward atomic exchange to be responsible for this phenomenon. The time required for complete homogenization of the alloys is 5-7 months.

Johnson and Christian [286] made a special study of the properties of germanium—silicon alloys. The constant of the cubic lattice varies almost linearly from 0.5426 for pure silicon to 0.5659 nm for pure germanium. Alloy density also varies monotonically from 6.323 for pure germanium to 2.328 g/cm^3 for pure silicon (Fig. 67).

Dismukes et al. [182] investigated the thermal conductivity, temperature function of thermal emf, electrical conductivity, and Hall current-carrier mobility in highly alloyed Ge—Si alloys with impurity concentrations of $2 \cdot 10^{18}$-$4 \cdot 10^{20}$ cm^{-3} over the temperature range 27-1027°C.

On the basis of the data obtained, it was established that alloys containing B (p-type) and P (n-type) have the optimum parameters for use in thermoelectric energy converters. The maximum value of the dimensionless coefficient of thermodynamic Q at 1200°K is 0.8 for the p-type alloy $Ge_{0.15}Si_{0.85}$ (with a hole concentration of $2.1 \cdot 10^{20}$ cm^{-3}) and 1.0 for the n-type alloy $Ge_{0.15}Si_{0.85}$ (with an electron concentration of $2.7 \cdot 10^{20}$ cm^{-3}).

The forbidden-zone width E_g for germanium—silicon alloys, determined from light absorbtion, increases from $1.12 \cdot 10^{-19}$ J (0.7 eV) for germanium to $1.92 \cdot 10^{-19}$ J (1.2 eV) for silicon, with an inflection in the zone-width curve at a content of about 12-15 at-% (Fig. 68). The forbidden-zone width thus varies with solicon content over two linear segments (the alloys with a silicon content of less than 12-15 at-% were monocrystals and those with a higher silicon content were polycrystalline). Levitas et al. [288] determined the forbidden-zone width in thoroughly homogenized germanium—silicon alloys from the slope of the curve representing the logarithm of resistance as a function of inverse absolute temperature in the intrinsic-conductivity region. According to their data, the forbidden-zone width rises abruptly from $1.18 \cdot 10^{-19}$ J (0.74 eV) for germanium to $1.79 \cdot 10^{-19}$ J (1.12 eV) at 50 at-% Si, this being the value for pure silicon. A slight maximum is observed at 75 at-% Si, but it must be attributed to experimental error. In contrast

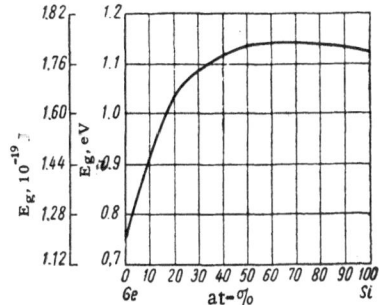

Fig. 68. Variation in forbidden-zone width E_g of germanium as a function of silicon content [288].

to the results of Johnson and Christian [286], Levitas et al. found the variation in forbidden-zone width with increasing silicon content to be nonlinear. Wang and Alexander [289, 290] investigated the hardness of germanium—silicon alloys. Their measurements showed that hardness varies continuously and almost linearly from germanium to silicon, which confirms the existence of a continuous series of solid solutions in the germanium—silicon system.

Sahm and Gnau [476] investigated the pressure-sintering of $Si_{0.63}Ge_{0.37}$ alloys containing phosphorus (n-type) and studied their thermoelectric properties. It was established that the n-type alloys were better compacted during sintering than the p-type alloys. The electrical properties of the pressure-sintering alloys differed from those of alloys produced by zone melting, but the differences almost completely disappeared after recrystallization. The pressure-sintered and recrystallized alloys had a density of 99.5%, a current-carrier concentration of $1.8 \cdot 10^{20}$, a thermal emf of 113 μV/deg, an electrical conductivity of $9.5 \cdot 10^{-4}$ ohm$^{-1} \cdot$ cm^{-1}, and a thermal conductivity of 0.050 W/cm·deg.

Tin — Germanium System. Gueriler and Pirani [291] employed thermal and microscopic analysis to study the phase diagram of the germanium—tin system at Ge contents between 0 and 40 at-%, using specimens prepared by fusion under a flux. They established that the solubility of germanium in solid tin is 0.1-0.01%, while tin is insoluble in germanium. The solidus temperature was found to be 230°C, while the liquidus temperature of alloys containing 15% Ge turned out to be 570°C. On the basis of these data, they concluded that the germanium—tin system produes a diagram of the simple eutectic type, with the melting point of the eutectic lying 1°C below that of tin. These conclusions were comfirmed in the more thorough investigation conducted by Stoehr and Klemm [263]. The phase diagram was worked out by thermal and x-ray analysis of vacuum-fused specimens (Fig. 69).

The solubility of tin in germanium at 195°C is less than 1 at-% and that of germanium in tin is less than 0.6 at-%. More recent investigations [292, 293] found the solubility of tin in solid germanium to be $2 \cdot 10^{18}$ atoms/cm^3. The tin enters into a solid solution and increases the lattice constant of the germanium [292, 295]. The variation in the lattice constant of Ge as a function of Sn content is shown below:

Sn content, at-%	0	0.12	0.66	1.04
Lattice constant, nm	0.5667	0.5657	0.5663	0.5666

Addition of germanium improves the mechanical properties of tin. Thus, the Brinell hardness of tin doubles at a Ge content of 0.35%, while its relative elongation decreases by 20% [191]. At 30°C, germanium present in white tin can serve as a nucleus for its conversion to gray tin [296].

Pokrovskii and Tissen [297] investigated the influence of addition of between 0.5 and 5 at-% Ge on the surface tension of tin over the temperature range 400-500°C. They established that less than 0.5 at-% Ge has virtually no effect on the density of tin but that the latter is somewhat reduced (by no more than 0.3%) by up to 2 at-% Ge.

Lead — Germanium System. The phase diagram of the Ge—Pb system is of the simple eutectic type and was worked out by thermal and microscopic analysis (Fig. 70) [298]. Later research [135] confirmed that the eutectic in this system lies near pure lead. According to Thurmond [163], the melting point of the eutectic is 0.8°C below that of lead.

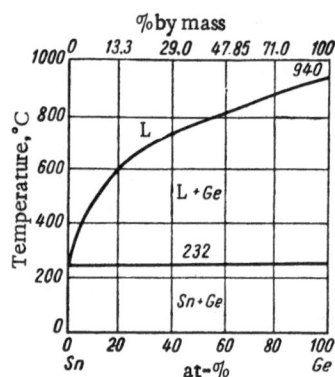

Fig. 69. Phase diagram of the tin—germanium system [291].

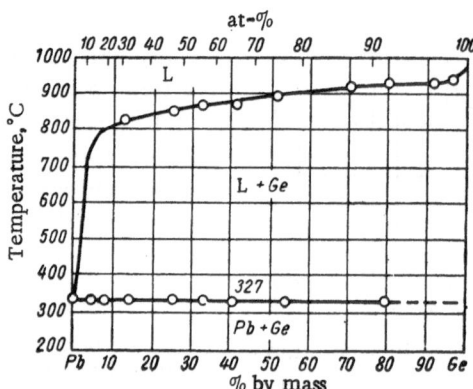

Fig. 70. Phase diagram of the lead-germanium system [289].

COMPOUNDS WITH NITROGEN-SUBGROUP ELEMENTS AND OTHER COMPOUNDS

Germanium Nitrides. The compounds Ge_3N_2 (11.4% N) and Ge_3N_4 (20.46% N) have been found in the germanium—nitrogen system. The nitride Ge_3N_2 was obtained by heating metallic germanium in flowing nitrogen at 800–950°C and by decomposition of GeNH or Ge_3N_4. The compound is a dark brown powder that sublimates at 650°C [299, 459].

The following methods were employed to produce the nitride Ge_3N_4:

a) passage of ammonia over metallic germanium heated to 650–700°C or over germanium dioxide at 700–750°C;

b) decomposition of the germanate Ge_2N_3H by heating to 300°C [300–302]. Attempts to prepare Ge_3N_4 from the elements at atmospheric or high pressure have been unsuccessful [3].

The nitride Ge_3N_4 has a rhombohedral structure of the phenacite type. Its lattice constants are a = 0.8587 nm and α = 1.88 rad (107° 48') [303]. According to other data, Ge_3N_4 has a rhombic structure with the lattice constants a = 1.384, b = 0.906, and c = 0.818 nm [305]. It is a colorless or light yellow substance with a reddish tint; when heated over the range 600–1000°C, it decomposes into the elements and a more volatile cinnamon-brown substance, probably GeN_2. On heating to 900°C, it decomposes into germanium and nitrogen [304]. The nitride Ge_3N_4 has a density of 5.25 g/cm³, a heat of formation $-\Delta H$ = 65.31 kJ/mole (15.6 kcal/mole) [3, 304], and an electrical resistance of more than 10^8 ohm·cm [77]. It is reduced to the metal and ammonia by hydrogen at 600–700°C and converted to oxides by oxygen at 600–900°C [306]. Its magnetic susceptibility is $3.3·10^{-7}$.

Germanium Arsenides. The germanium—arsenic system has been found to contain the compounds $GeAs_2$ (32.65% Ge) and GeAs (49.21% Ge), which melt congruently at 732 and 737°C respectively and are homogeneous over broad composition regions. The phase diagram, which is given in Fig. 71, was worked out by thermal, x-ray, microscopic, and magnetic analysis [264].

The compound $GeAs_2$ forms a eutectic with the α-solid solution of germanium in arsenic, while the compound GeAs forms one with the β-solid solution of arsenic in germanium. The hexagonal lattice of the α-solid solution of germanium in arsenic has constants that vary as the germanium content is increased from 0 to 15.5%, a falling from 0.3761 to 0.3708 nm and c rising from 1.0541 to 1.0731 nm.

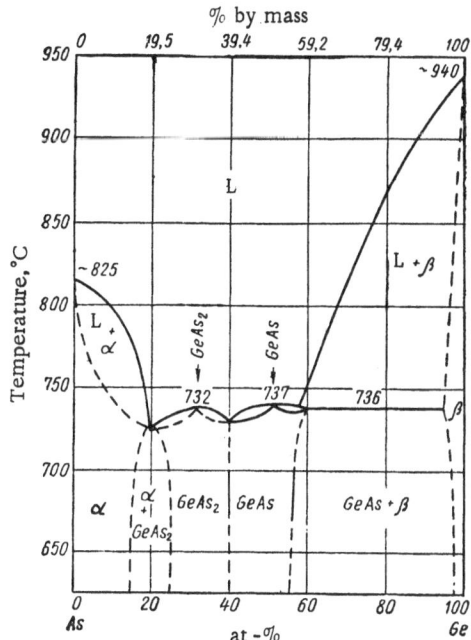

Fig. 71. Phase diagram of the arsenic−germanium system [264].

The lattice constant of germanium remains almost unchanged when about 3.1% As is dissolved in it [264].

The compound $GeAs_2$ has a rhombic structure with the lattice constants a = 0.3721, b = 1.0120, and c = 1.4739 nm. According to Bryden [307], this compound has a rhombic lattice (Pbam space group) with the constants a = 1.476, b = 1.016, and c = 0.3728 nm. The elementary cell contains eight formal units.

The compound GeAs has a monoclinic structure with the lattice constants a = 2.2124, b = 0.3777, and c = 0.9448 nm; β = 0.77 rad (43.97°) [308].

The existence of the chemical compound GeAs in the germanium−arsenic system and the fact that it forms solid solutions with germanium were confirmed by an investigation of alloys containing up to 42.4% As prepared by electrolysis. A melt consisting of borax, germanium dioxide, and sodium arsenate was electrolyzed at 800°C [284]. The density of the alloys in the system increased with rising arsenic content:

As content, %	0	1.4	2.3	8.6	50.0	67.0	72.0	86.5	100.0
Density, g/cm³	5.323	5.317	5.388	5.406	5.371	5,389	5.435	5.636	5.73

The solubility of germanium in arsenic at 685°C is 15.5% (16 at-%), while that of arsenic in germanium is 3.1% (3 at-%). The heat of solution of arsenic in solid germanium is ΔH = 30.38 kJ/mole (7.4 kcal/mole) and its temperature coefficient $\Delta H/T$ = 10.88 J/mole·deg (2.6 cal/mole·deg) [251].

Addition of germanium greatly improves the stability of arsenic when it is stored in air [264].

Germanium Antimonides. Thermal and microscopic analysis [131] established that germanium and antimony form a system of the simple eutectic type with a eutectic point at a Ge content of 12.3% (19 at-%) and a temperature of 588°C.

Glasov and Chizhevskaya [309] used the viscosity method to determine the position of the liquidus for this system. Their results were in almost complete agreement with the curve constructed by thermal analysis [135].

Stoehr and Klemm conducted more careful and thorough investigations [264]. On the basis of their data, the position of the eutectic point was refined and it was established that germanium and antimony have limited mutual solubility in the solid state. They found that the eutectic point lies at a Ge content of 8.53% (13.5 at-%) and a temperature of 592°C. Figure 72 shows a diagram constructed from their data [264]. The dash line represents the position of the liquidus and eutectic point determined from the data of Ruttewit and Masing [135]. On entering into a solid solution with antimony, germanium undergoes a change in lattice constants. The crystal structure is shown in Table 12.

The diagram in Fig. 72 is in poor agreement with the data of Thurmond and Struthers [164], according to whom the solubility of antimony in germanium is 0.02, 0.003, and $9.0 \cdot 10^{-5}$ at-% at temperatures of 880, 730, and 580°C respectively.

Table 12. Crystal Structure of Ge−Sb Alloys

Metal	Type and lattice constants					
	Hexagonal			Rhombohedral		
					α	
	a, nm	c, nm	c/a	c, nm	rad	deg
Pure antimony	0.4296	1.124	2.617	0.6217	1.53	87.4
Solution of Ge, Sb	0.4266	1.130	2.649	0.6201	1.521	86.9

Bismuth − Germanium System. Ruttewit and Masing [135] employed thermal and microscopic analysis to investigate the germanium−bismuth system. They established that its phase diagram is of the eutectic type, with the eutectic consisting of almost pure bismuth (Fig. 73). The two elements are mutually insoluble in the solid state. The character of the diagram was confirmed by Stoehr and Klemm [264], but these authors established that bismuth and germanium are mutually soluble in the solid state, the solubilities at 250°C amounting to 0.52% by mass (1.5 at-%) for germanium in bismuth and 5.5% by mass (2 at-%) for bismuth in germanium. The melting point of the eutectic was found to be 271°C in both investigations. According to Thurmond [163], the melting point of the eutectic is 0.1°C below that of bismuth and its germanium content is 0.1 at-%.

Germanium Oxides. Two compounds are known in the germanium−oxygen system: the monoxide GeO (18.06% O) and the dioxide GeO_2 (30.59% O). Germanium monoxide can be produced by sublimation of a mixture of germanium dioxide and metallic germanium in a vacuum or by heating germanium in a nitrogen atmosphere containing an admixture of oxygen or in flowing CO_2 at 800-900°C. In the first procedure, formation of GeO begins at 600°C and proceeds rapidly at 1000°C [3, 5, 311, 313, 314]. Germanium dioxide can be obtained by heating the metal in oxygen or by oxidation of its sulfides.

Hoch and Johnston [315] were the first to conduct systematic investigations of the germanium−oxygen system, employing high-temperature x-ray diffraction analysis and electrical-conductivity measurements. They established that GeO is not formed when an equimolecular mixture of germanium and GeO_2 is heated at 750-1400°C. X-ray patterns obtained at 850-930°C exhibit only germanium lines. At 1000°C and above, the germanium lines disappear from the

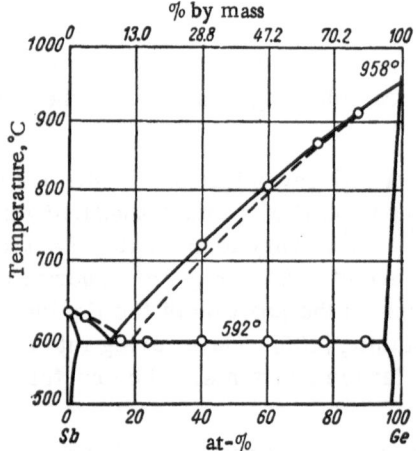

Fig. 72. Phase diagram of the anti-
mony−germanium system [264].

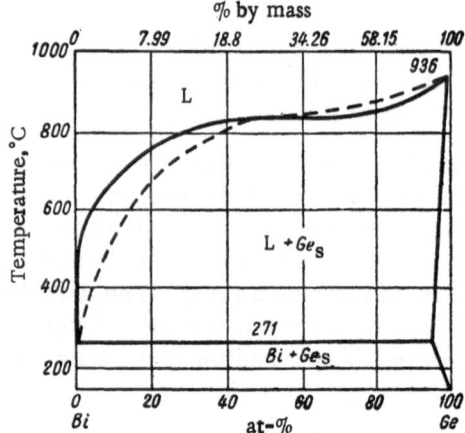

Fig. 73. Phase diagram of the bismuth−
germanium system [135].

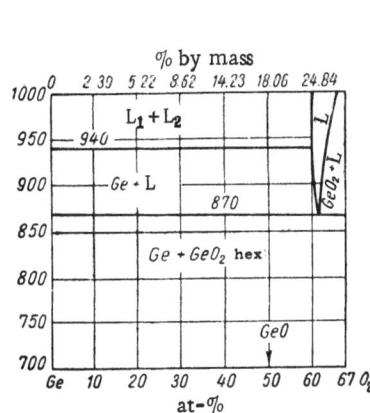

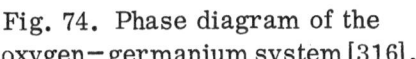

Fig. 74. Phase diagram of the
oxygen—germanium system [316].

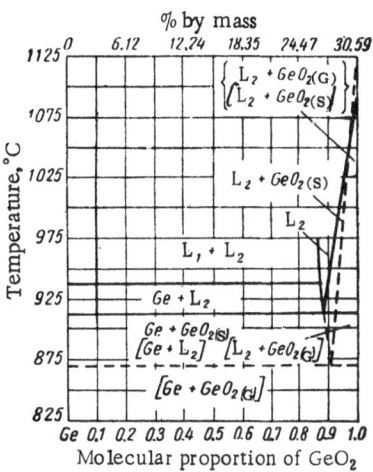

Fig. 75. Phase diagram of the
Ge—GeO₂ system [320].

patterns and a liquid-like structure is observed, although the specimen remains solid; this indicates disordering of the germanium lattice, which occurs at 965°C. Germanium lines reappear in the x-ray patterns when the mixture is cooled to 930°C. The authors attribute the absence of GeO_2 lines in patterns obtained above 750°C to the high solubility of oxygen in germanium, which amounts to 60 at-%. On the basis of these data, they concluded that germanium monoxide does not have a stable region in the germanium—oxygen system.

A second investigation of the germanium—oxygen system was conducted by Candidus and Tuomi [316], who employed x-ray diffraction analysis.

The data obtained by these investigators essentially confirmed the results of prior research, as can be seen from the diagram in Fig. 74. They also failed to detect any stability region for germanium monoxide in the system. Only two phases (germanium and the hexagonal modification of GeO_2) were found at oxygen contents between O and 67 at-% below 870°C. Two phases (solid germanium and a liquid containing 60 at-% Ge) were also observed at temperatures of 870-940°C and an oxygen content of up to 60 at-%. The tetragonal modification of germanium dioxide was found to be in equilibrium with the eutectic liquid at a higher oxygen content. The system contained two liquid phases (germanium and a liquid with the composition Ge_2O_3 [24.84% O]) at oxygen contents of from O to 60 at-% above 940°C.

The existence of the compound GeO and the occurrence of a polymorphic transformation in GeO_2 were established by various authors [317-319]. Trumbore et al. [320] employed chemical, microscopic, and x-ray analysis to work out the phase diagram of the oxygen—germanium system in the Ge—GeO₂ region (Fig. 75) and found that the eutectic in this system lies at 912°C.

Germanium Sulfides. The germanium—sulfur system has been found to contain two sulfides: GeS_2 (53.10% Ge) and GeS (69.36% Ge). The compound GeS_2 has a rhombic structure of the GeS_2 type with the lattice constants a = 1.1683, b = 2.2385, and c = 0.6783 nm [321, 322]. It is formed when GeO_2 is heated in sulfur vapor, when a germanium oxide is precipitated from concentrated acid solutions with hydrogen sulfide, or when powdered germanium is exposed to a mixture of H_2S and sulfur vapor at 850°C [3, 323-325]. The density of GeS_2 is 2.7 g/cm³ (2.942 g/cm³ according to other data [3, 326]), its melting point is 800°C, and its Mohs hardness is 2-2.5. The equilibrium vapor pressure over solid GeS_2 ranges from 133.3 MN/m² to 399.9 N/m² (from 10^{-3} to 3 mm Hg) over the temperature interval 440-730°C. The extent to which this compound dissociates into GeS and S_2 increases with temperature and complete dissociation occurs at about 800°C. The vapor tensions of the dissociation products (S_2, GeS, and a small amount of S_4) at different temperatures are given below, being based on the data of Kenworthy et al. [325]:

Temperature,°C	425	450	475	500	525	550
Total vapor tension						
N/m^2	$7.33 \cdot 10^{-4}$	$4.93 \cdot 10^{-3}$	$1.76 \cdot 10^{-2}$	$7.91 \cdot 10^{-2}$	$1.36 \cdot 10^{-1}$	$2.54 \cdot 10^{-1}$
mm Hg	$5.5 \cdot 10^{-6}$	$3.7 \cdot 10^{-5}$	$1.25 \cdot 10^{-4}$	$5.94 \cdot 10^{-4}$	$1.02 \cdot 10^{-3}$	$1.91 \cdot 10^{-3}$

Gurovich [327] produced germanium disulfide (GeS_2) by precipitation from 6 N sulfuric acid solution with hydrogen sulfide. The resultant digermanide contained 53.05% Ge (theoretical content, 53.1%) and 47.45% S (theoretical content, 46.9%). Analysis of the compound for germanium dioxide showed it to be absent.

Germanium sulfide (GeS) exists in two modifications, amorphous and crystalline. Crystalline GeS has a rhombic structure of the GeS type with the lattice constants a = 0.4298, b = 1.0440, and c = 0.3647 nm [411]. This compound can be produced by heating a mixture of GeS_2 and germanium, as well as by heating GeS_2 in flowing hydrogen or by sublimating germanium in flowing H_2S at 850°C [3, 325]. The amorphous modification of GeS is yellowish red, while the crystalline modification is black. Their densities are 3.31 and 4.012 g/cm^3 respectively. The compound has a melting point of 530°C; it vaporizes above 650°C and completely dissociates at 1400–1500°C. In a vacuum, it readily sublimates at 305–376°C. Its heat of sublimation at 298°K is ΔH = 171.24 J/mole (40.9 kcal/mole) and its heat of formation is ΔH = 108.02 kJ/mole (25.8 kcal/mole) [326].

The crystalline modification of GeS has a Mohs hardness of 2. The molecular heat capacity of gaseous GeS over the temperature range 25–427°C is defined by the equation [326]

$$C_p = 8.873 - 7.394 \cdot 10^{-4} \cdot T^{-2}.$$

Yabumoto [331] investigated the electrical and optical properties of GeS. He established that its electrical resistance is an exponential function of temperature. The activation energy E_{act} is $(1.18–1.6) \cdot 10^{-19}$ J (0.74–1.0 eV) for high temperatures and $(0.4–0.8) \cdot 10^{-19}$ J (0.25–0.5 eV) for low temperatures. The photoconductivity maximum occurs at 0.8μ (0.7μ in a vacuum) at room temperature. The compound GeS has p-type conductivity at high temperatures; its thermal emf is about 1 $\mu V/deg$.

Germanium Selenides. The compounds $GeSe_2$ (31.50% Ge) and GeSe (47.90% Ge) were produced by fusing a mixture of germanium and selenium powders in a test tube made of high-melting glass, which was preliminarily filled with carbon dioxide. The same compounds can also be obtained by exposing germanium chloride to hydrogen selenide.

The monoselenide GeSe has a tetragonal or psuedotetragonal structure with the lattice constants a = 0.8847, c = 0.9779 nm, and c/a = 1.105. It is a hard mass with a color reminiscent of that of galena. Its density at 25°C is 5.30 g/cm^3 and its Mohs hardness does not exceed 2. Its melting point is 667°C and it readily vaporizes above this temperature. Selenium slowly burns when heated in air at 460–470°C [332].

The investigation of germanium selenide conducted by Liu Ch'iun-Hua et al. [334] showed that it has a deformed rhombic lattice of the sodium chloride type and is isostructural with germanium and tin sulfides. Its lattice constants (D_{2h}^6-Pbmn space group) are a = 0.4383, b = 0.3832, and c = 1.0821 nm. Data indicating that germanium selenide has a tetragonal structure [332] are obviously incorrect.

Germanium diselenide, $GeSe_2$, has a rhombic structure with the lattice constants a = 1.2986, b = 0.6943, and c = 2.2134 nm [332]. It is yellow in color, sometimes with an orange tint. Its

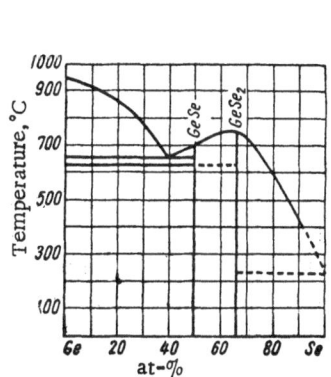

Fig. 76. Phase diagram of
the selenium—germanium
system [333].

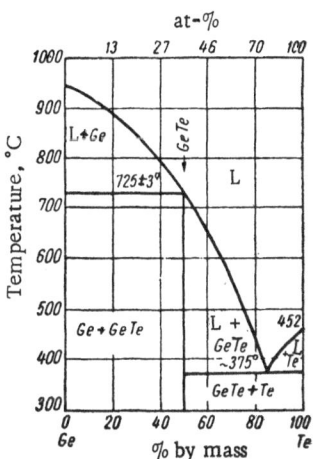

Fig. 77. Phase diagram of
the tellurium—germanium
system [335].

density at 25°C is 4.56 g/cm^3 and its Mohs hardness does not exceed 2. Its melting point is 707°C
and it readily vaporizes above this temperature.

Liu Ch'iun-Hua et al. [333] employed thermal and x-ray analysis to work out the phase diagram of the germanium—selenium system, which is shown in Fig. 76. They confirmed the existence of the compounds GeSe and GeSe$_2$; in contrast to the data of Ivanov-Émin [332], they found
the diselenide lattice to have the constants a = 0.6953, b = 1.2220, and c = 2.3036 nm.

The monoselenide decomposes by the peritectic reaction GeSe $\rightleftharpoons$ melt + GeSe$_2$ at 670°C
and forms a eutectic with germanium at 650°C at 40 at-% Se. It undergoes a polymorphic transformation at 620°C, as can readily be seen in the cooling curves for Ge—GeSe alloys.

Germanium diselenide melts congruently at 740°C. The two-phase GeSe$_2$—Se region has
been the subject of little research. The authors assume that this part of the system contains a
degenerate eutectic, which is virtually indistinguishable in composition and melting point from
pure selenium.

Germanium Tellurides. The phase diagram of the germanium—tellurium system,
which is shown in Fig. 77, was worked out by Klemm and Frischmuth [335], who employed thermal and x-ray analysis. The compound GeTe (63.74% Te) has been found in this system, being
formed by a peritectic reaction at 725°C. It produces a eutectic with tellurium at 375°C and a
Te content of 91%.

A germanium—tellurium alloy with the composition Te$_{49.2}$Ge$_{50.8}$ (63.0% Te), i.e., close to
the composition of the compound GeTe [336], has a face-centered rhombohedral structure (te-
tragonally distorted NaCl structure) with the lattice constants a = 0.5980 nm and α = 1.546 rad
(88.35°). The value of a rises as the temperature is increased.

Shelimova et al. [337] employed thermal analysis, microstructural analysis, microhardness measurements, and thermal-expansion measurements to investigate the homogeneity region based on the compound GeTe and the variation in the temperature at which GeTe is converted
from the rhombohedral to the cubic structure as a function of composition. Alloys containing
from 49 to 52 at-% Te were prepared by fusing the high-purity elements (99.996% Te and Ge
with ρ = 40 ohm·cm) in evacuated quartz ampules. After fusion, the melt was carefully mixed
and cooled in air; the alloys were then annealed at different temperatures and holding times in
an argon atmosphere and quenched in ice water.

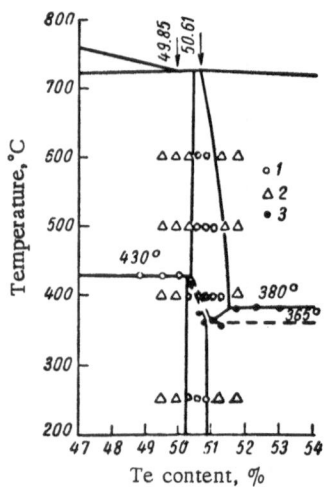

Fig. 78. Phase diagram of
the Ge−Te system in the vicinity of the compound GeTe
[337]. 1) Monophasic alloys;
2) diphasic alloys; 3) thermal-
analysis data.

The data yielded by thermal and microstructural analysis and thermal-expansion measurements were used to construct a phase diagram for the germanium−tellurium system
in the vicinity of the compound GeTe (Fig. 78). It was established that GeTe melts with an open maximum and is not
formed by a peritectic reaction, as was thought by Klemm and
Frischmuth [335]. This conclusion is in agreement with the
data of McHugh and Tiller [338]. The homogeneity region
based on the high-temperature modification of GeTe lies between 50.3 and 51.5 at-% Te (430°C), while that based on the
low-temperature modification is somewhat narrower (extending from 50.2 to 50.9 at-% Te). The polymorphic-transformation temperature is 430°C on the germanium side and 365°C on
the tellurium side. The GeTe + Te eutectic has a melting
point of 380°C.

According to Schubert and Fricke [339], the compound
TeGe exists in two modifications and the structure described
above has a low-temperature modification. The high-temperature modification has a face-centered structure of the NaCl
type with a lattice constant a of 0.601 nm at 460°C. The experimental density of the compound is 6.200 g/cm^3 and its x-ray
density is 6.290 g/cm^3 [336]. The resistivity of a TeGe spe-
cimen of stoichiometric composition, prepared from germanium with a resistance of several
ohm·cm and 99.998% pure tellurium is 10^{-5} ohm·cm. The compound TeGe has pronounced semi-
metallic properties and a positive temperature coefficient of electrical resistance [79].

Investigation of the electrical properties of TeGe [78, 80] has shown that this compound
exhibits p–type conductivity; the current-carrier concentration is about $9 \cdot 10^{20}$ cm^{-3}, regardless
of the preparation technique. There is no information in the literature on the nature of conductivity in GeTe. According to Moriguchi and Koga [79], GeTe is a semimetal with overlapping
valence and conduction zones. The high current-carrier concentration is due to the nonstoichiometric composition of GeTe [78, 80]. More recently, Kolomoets [340] investigated the electrical properties of GeTe in order to determine the nature of its conductibity and confirmed previous conclusions to the effect that the high current-carrier concentration results from the
nonstoichiometric composition of the compound [78, 80].

Further research [154, 310, 439] made it possible to establish that solid germanium telluride is a semiconductor, its high charge-carrier concentration being due to its deviation from
stoichiometry. Moreover, an investigation of the thermal emf of this compound in the solid and
liquid states [439] demonstrated that the emf is reduced by a factor of six when the solid melts
(α_{GeTe_S} = + 130 μV/deg, while α_{GeTe_L} = + 21 μV/deg). The sign of the emf remains the same,
however, which indicates that GeTe remains semiconductive after melting. Figure 79 shows the
thermal emf of germanium telluride as a function of temperature in the solid and liquid states.

The substantial decrease in thermal emf on melting results from equalization of the electron and hole mobilities. The presence of hole conductibity and the decrease in thermal emf
when the melt temperature is raised confirm that germanium telluride is a semiconductor.

The compound TeGe does not exhibit superconductivity at least down to a temperature of
1.02°K [12].

Andreev et al. [341] investigated the influence of high pressure on the electrical properties of germanium telluride (electrical conductivity, thermal emf, and Hall constant). The measurements were made with monocrystalline specimens at a pressure of 100,000 MN/m^2 (10,000

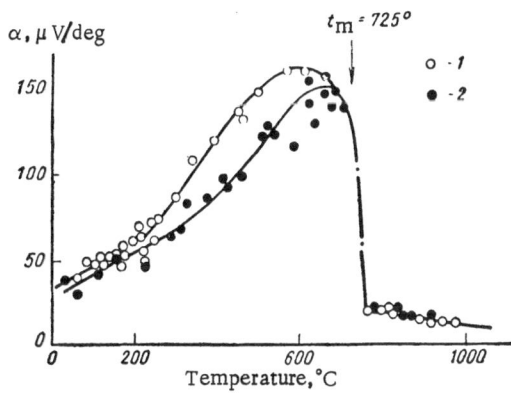

Fig. 79. Thermal emf of germanium telluride in the solid and liquid states as a function of temperature [439]. 1) Purified TA1 tellurium; 2) unpurified TA1 tellurium.

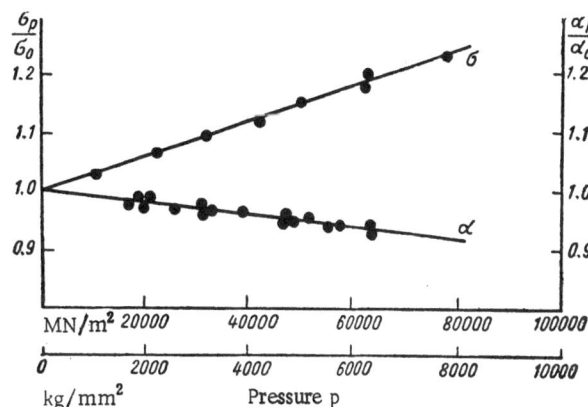

Fig. 80. Electrical conductivity and thermal emf of GeTe as a function of pressure at at 27°C [341].

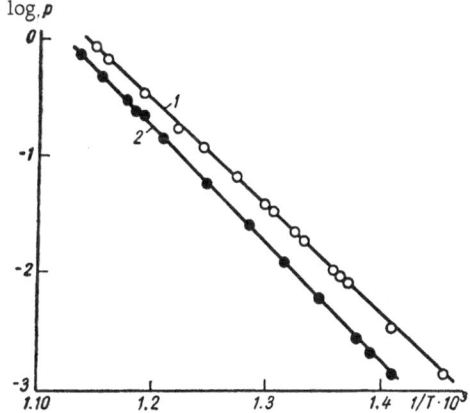

Fig. 81. Variation of vapor saturation-pressure with temperature for selenides (1) and tellurides (2) of germanium.

kg/mm^2) over the temperature range 20-200°C. This investigation established that, as the pressure is raised, the electrical conductivity increases linearly, the thermal emf drops, and the Hall constant remains unchanged. Figure 80 shows the relative change in electrical conductivity and thermal emf at 27°C, expressed as a percentage per 10,000 MN/m^2 (1000 kg/mm^2).

The authors consider this relationship to result from the change in effective mass as a function of pressure.

Liu Ch'iun-Hua et al. [334] determined the saturated vapor pressure of solid germanium selenide and germanium telluride. The telluride and selenide were synthesized by fusing the components in sealed quartz ampules evacuated to 133.3-13.33 mN/m^2 (10^{-3}-10^{-4} mm Hg). The ampules were slowly heated to 1000°C and held at this temperature for 1-2 h.

The saturated vapor pressure of the germanium selenide and telluride was measured by Knudsen's method [342]. Fig. 81 shows this pressure as a function of temperature. In accordance with the equations obtained, the heats of sublimation of GeSe and GeTe are ΔH°_{778} = 179.6 kJ/mole (42.9 kcal/mole) and ΔH°_{794} = 191.7 kJ/mole (45.8 kcal/mole) respectively.

An electron-diffraction study of the germanium−tellurium system [240] established the existence of the new phase Ge$_2$Te$_3$, which is a superstructure with a hexagonal elementary cell (D$^3_{3d}$ space group) and the lattice constants a = 0.432 and c = 5.30 nm.

CHAPTER 6

MULTICOMPONENT GERMANIUM—CONTAINING ALLOYS

SYSTEMS IN WHICH NO CHEMICAL COMPOUNDS ARE FORMED

Nitrogen – Manganese – Germanium. Alloys in the ternary N–Mn–Ge system were prepared by fusion of the components in aluminum oxide crucibles and, after powdering, were nitrided at 600°C in a flowing ammonia–hydrogen mixture [343]. The nitrided alloys (with the composition Mn_3Ge) exhibited a hexagonal phase with the lattice constants a = 0.560, c = 0.452 nm, and c/a = 0.806.

The authors believe that this phase is the known phase $Mn_{3.25}Ge$, the increase of 5% in its lattice constants being due to solution of the nitrogen. No cubic nitrides were found in the system.

Nitrogen – Nickel – Germanium. According to x-ray data [343], a nitrided alloy with the composition $Ni_{85}Ge_{15}$ contained two phases with the lattice constants a = 0.3556 nm and a = 0.3575 nm. An alloy with the composition Ni_3Ge exhibited a face-centered cubic phase with the lattice constant a = 0.3580 nm. No ternary compounds were detected in the system.

Aluminum – Magnesium – Germanium. Badaeva and Kuznetsova [344] investigated the Al–Mg–Ge system by microscopic and thermal analysis of alloys prepared by fusing high-purity aluminum with magnesium–germanium alloys in corundum crucibles under a KCl–LiCl flux. Diagrams were constructed for the quasibinary Al_3Mg_2–Mg_2Ge and Al–Mg_2Ge segments and for the liquidus of the Al–Mg–Ge system (Figs. 82-84). These authors established that Mg_2Ge produces diagrams of the eutectic type with aluminum and with the compound Al_3Mg_2. Eutectic equilibrium is observed in the quasibinary Al–Mg_2Ge segment at 624°C. A singular syncline is observed in the primary-crystallization region, indicating the presence of molecular structures of Mg_2Ge in the aluminum-based solution.

Aluminum – Antimony – Germanium. The ternary Al–Sb–Ge system has been the subject of a number of investigations [345-349] conducted to study the interaction among the alloying elements and to determine the solubility of antimony, aluminum, or AlSb in the germanium lattice. Thus, Glazov and Liu Ch'iun-Hua [346] established that the maximum solubility of AlSb in germanium is 1.5 mol-%. Other investigations [345, 347] showed that the mutual solubility of aluminum and antimony (1:1) at 820°C is 3% by mass [345] or 2.81 at-% [347]. Figure 85 shows the variation in the solubility of aluminum and antimony in germanium as a function of temperature.

The character of the heterogeneous equilibrium between the solid and liquid phases when the latter was pulled from a melt alloyed with both aluminum and antimony was studied by Zemskov et al. [350], who employed alloys containing 0.065-1.00 at-% Al, 1.935-1.00 at-% AlSb, and 98 at-% Ge and alloys containing 0.097-1.500 at-% Al, 2.903-1.500 at-% Sb, and 97 at-% Ge.

These authors established that the solid phase crystallizes with an equiatomic Al:Sb ratio when the ratio of the number of Sb atoms to the number of Al atoms in the melt is 22-25, i.e.,

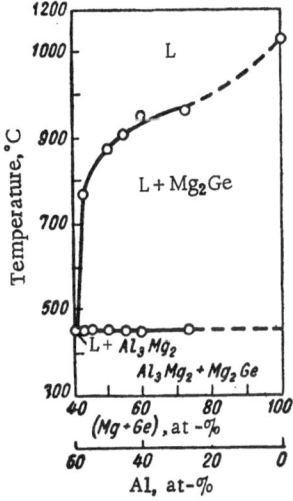

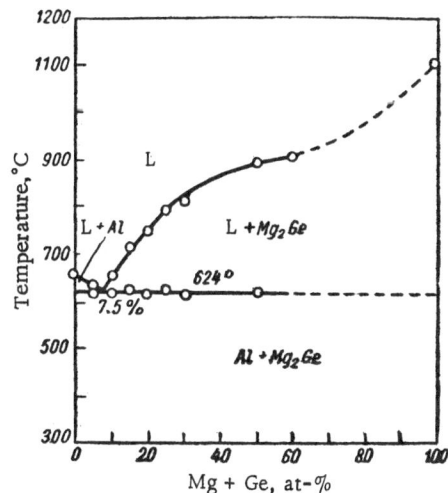

Fig. 82. Diagram of quasibinary $Al_3Mg_2 - Mg_2Ge$ segment of the Al−Mg−Ge system [344].

Fig. 83. Diagram of quasibinary Al−Mg_2Ge sigment of the Al − Mg − Ge system [344].

close to the distribution-coefficient ratio for Al in Sb for binary germanium-containing systems ($2.9 \cdot 10^{-3}$ for Sb and $7 \cdot 10^{-2}$ for Al). The observed behavior of Al and Sb in germanium is analogous to that of In in Sb [351] and can be attributed to dissociation of the antimonides in the melt.

The Ge−AlSb segment in the Al−Sb−Ge system is not quasibinary.

Aluminum − Phosphorus − Germanium. Abrikosov et al. [352] studied this system with ternary and binary germanium alloys having aluminum and phosphorus contents of up to 3.0 at-% in order to establish the maximum, separate, and joint solubilities of these elements in germanium.

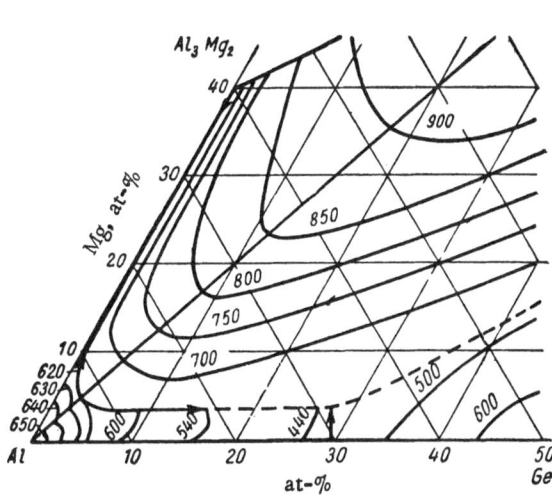

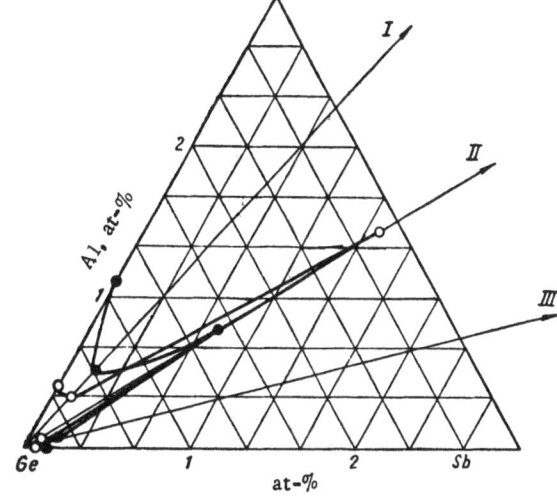

Fig. 84. Projection of the liquidus surface of the $Al-Mg_2Ge$ system [344].

Fig. 85. Isotherms for solubility of aluminum and antimony in germanium at Al:Sb ratios of 3:1 (I), 1:1 (II), and 1:3 (III) [345, 347].

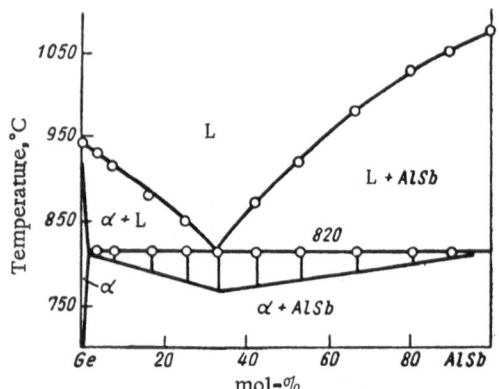

Fig. 86. Fusibility diagram of the Ge−
AlSb system [348].

Microscopic analysis and microhardness measurements demonstrated that the maximum solubility of phosphorus in germanium at 600°C is about 0.45 at-%, while that of aluminum in germanium at 470°C is about 1.2 at-% [347].

Abrikosov et al. [352] found that addition of aluminum to such an alloy leads to a significant increase in the solubility of phosphorus. Thus, the solubility of this element at 800°C is increased by a factor of five in the presence of an equiatomic amount of aluminum. The solubility of aluminum is also markedly improved in the presence of phosphorus.

Figure 87. shows isotherms for the joint solubility of aluminum and phosphorus in germanium, from which it can be seen that these elements have maximum solubility in the Ge−AlP segment.

The fact that the joint solubility of aluminum and phosphorus higher than the individual solubility of either element is attributed to formation of the chemical compound AlP in the germanium-based solution; this compound is very similar in crystal-chemical properties to the solvent.

Vanadium − Silicon − Germanium. The V_3Si-V_3Ge system has been studied by Holleck et al. [217] and Savitskii et al. [353], who employed microstructural and x-ray analysis and microhardness measurements of alloys prepared by fusion of the high-purity elements 99.8% V, 99.8% Si, and 99.9% Ge) in a purified helium atmosphere at a pressure of 70 kN/m² (0.7 atm) in an arc furnace with a nonreactive tungsten electrode. In order to even out their composition, the ingots were subjected to four remeltings and then homogenized in evacuated quartz ampules at 800°C for 2500 h.

Holleck et al. [217] prepared their alloys by hot pressing of a mixture of the powdered metals and subsequent homogenization in an argon atmosphere. Vanadium-rich alloys were annealed at 1300°C for 12 h, while vanadium-poor alloys were annealed at 1000°C for the same length of time. X-ray data were employed to construct the phase diagram of the V−Si−Ge system shown in Fig. 88. It was established that a continuous series of solid solutions is formed between the compounds V_3Si and V_3Ge, or, more precisely, that there is isomorphic substitution of silicon atoms for the germanium atoms in the V_3Si lattice.

Figure 89b and c show the variation in the lattice constant and microhardness of the $V_3(Si, Ge)$ solid solution. The T_k of ternary alloys containing 5, 10, 15, and 20 at-% Ge is 11.8, 8.7, 7.2, and 6.1°K respectively. Figure 89a shows the change in the T_k of $V_3(Si, Ge)$ solid solutions as the silicon is replaced by germanium.

All alloys in the V_3Si-V_3Ge system exhibit superconductivity; the transition temperature varies smoothly over the V_3Si-V_3Ge segment.

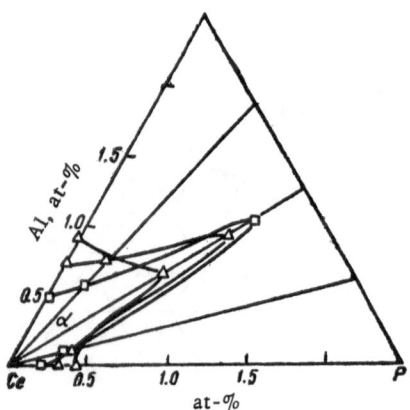

Fig. 87. Isotherms for joint solubility of aluminum and phosphorus in germanium [352].

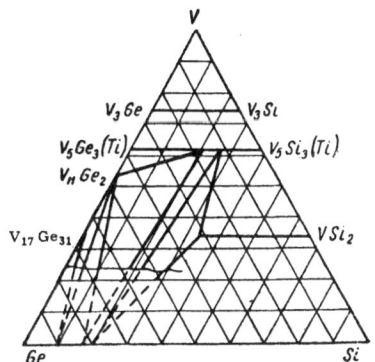

Fig. 88. Phase diagram of the V−Si−Ge system [217].

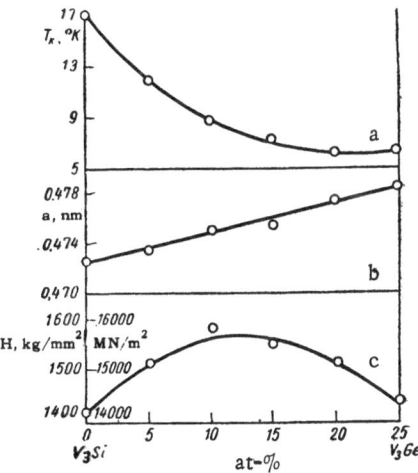

Fig. 89. Variation in T_k (a), lattice constant (b), and microhardness (c) of V_3(Si, Ge) solid solution [353].

Gallium (Indium) − Phosphorus − Germanium. Microscopic analysis and microhardness measurements of germanium alloys with gallium, indium, and phosphorus contents of up to 2.5 at-% were employed to investigate the joint solubility of these elements in germanium [354]. It was established that the maximum solubility of gallium phosphide at 800°C is about 1.1 at-% and that of indium phosphide at about 700°C is about 0.3 at-%. It was also noted that the solubility of indium is somewhat higher at all temperatures in the presence of an equiatomic amount of phosphorus than in the binary system, while the solubility of phosphorus is somewhat reduced. A different pattern is observed in the Ga−P−Ge system. The solubility of phosphorus at 800°C is markedly higher in the presence of an equiatomic amount of gallium than in the binary system, while the solubility of gallium is reduced.

Figures 90 and 91 show isotherms for the solubility of gallium, indium, and phosphorus in germanium, from which it can be seen that solubility is not maximal in the GaP−Ge and InP−Ge segments, which may be due to dissociation of the gallium and indium phosphides dissolved in the germanium.

Gold − Antimony − Germanium. Zwingmann [355] studied the ternary Au−Sb−Ge system with alloys prepared by fusion of the high-purity elements (99.95% Au, 99.95% Sb, and 99.9% Ge) in a hydrogen atmosphere in an electric furnace. Microscopic and thermal analysis was used to determine the liquidus surface of the ternary system and to construct the following polythermal-segment diagrams: a) with a constant Sb:Ge ratio of 9:1; b) $AuSb_2$− Ge; c) with a constant Ge content of 15 at-%

Figure 92 shows the ternary phase diagram of the Au−Sb−Ge system, from which it can be seen that the four-phase equilibrium L (35 at-% Au, 51 at-% Sb, 14 at-% Ge) + Sb = $AuSb_2$ + Ge and L (68 at-% Au, 17 at-% Sb, 15 at-% Ge) = Au + $AuSb_2$ + Ge occurs at 430 and 288°C.

Indium − Antimony − Germanium. The In−Sb−Ge system was investigated by microscopic and x-ray analysis and electrical-resistance measurements [356]. These data were employed to work out a phase diagram for the Ge−InSb system, which is of the eutectic type (Fig. 93). The eutectic contains 5 at-% germanium and melts at 512.5°C. A solubility study showed that InSb is insoluble in solid germanium, while about 0.75 at-% Ge dissolves in InSb.

Zhurkin et al. [351] investigated the quasibinary Ge−InSb segment. They found that liquid phases corresponding in composition to this segment are in equilibrium with solid phases whose compositions do not correspond to it and are displaced toward the antimony side. In order for the composition of the solid phase to correspond to that of the quasibinary segment, the composition of the liquid phase must be displaced toward the indium side; the ratio of indium to antimony in the melt should be slightly above 2.5.

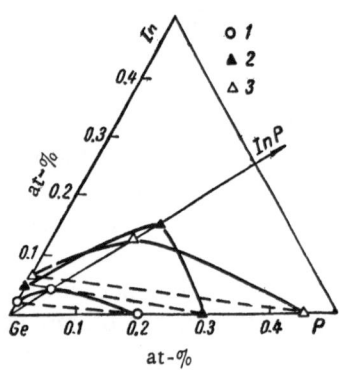

Fig. 90. Isotherms for solubility of indium and phophorus in germanium [354] at temperatures of: 1) 800°C; 2) 700°C; 600°C.

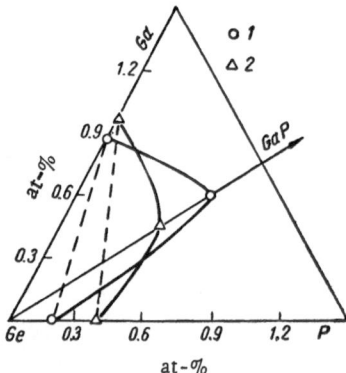

Fig. 91. Isotherms for solubility of gallium and phosphorus in germanium [354] at temperatures of: 1) 800°C; 2) 600°C.

These authors noted that the behavior of indium and antimony during crystallization of germanium melts with a high content of these elements is the same as when their contents are low. When the donor- and acceptor-impurity concentrations are small and the laws for dilute solutions hold, the interaction between the group-III and group-V atoms can be neglected and impurity trapping during crystallization is characterized by the distribution coefficients $K_{In} = 0.001$ and $K_{Sb} = 0.003$. The values are in good agreement with the In:Sb ratio mentioned above. This analogy serves to prove that the compound InSb dissociates above 620°C in molten germanium, which is in good agreement with the data on the thermal stability of InSb [351, 357].

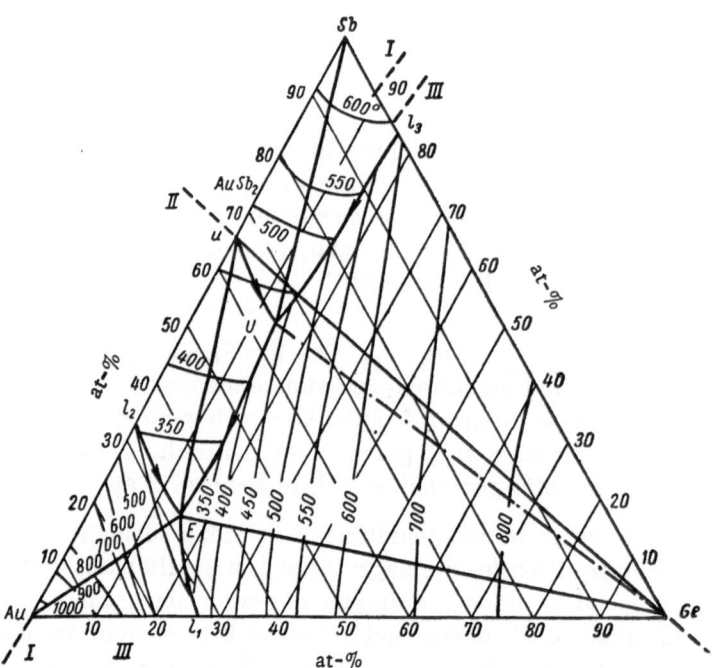

Fig. 92. Phase diagram of the Au–Sb–Ge system [355].

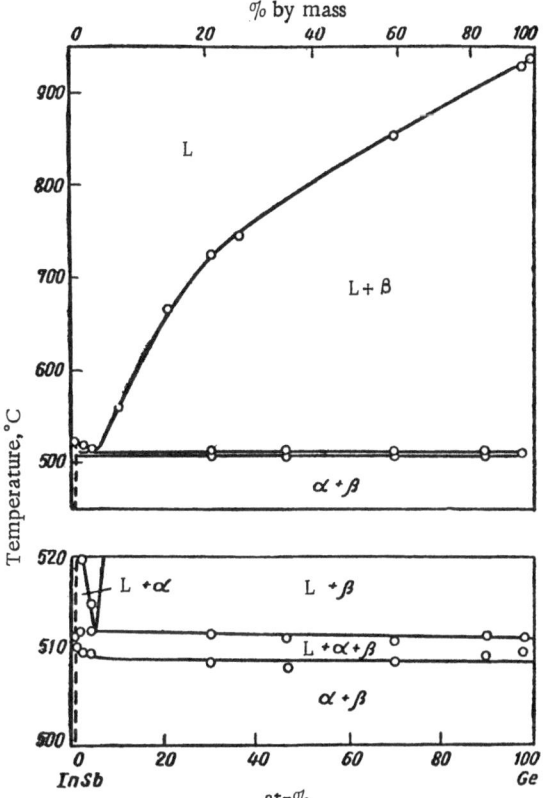

Fig. 93. Phase diagram of the Ge−In−Sb
system [356].

Indium − Tellurium − Germanium.
Wooley [446] used x-ray diffraction analysis to
study the phase composition of alloys in the
GeTe−InTe segment and alloys with compositions
lying in the GeTe−In_2Te_3−InTe−Ge region. The
alloys were quenched from the melt region and
some were annealed at 400°C. Figure 94 shows
a phase diagram for GeTe−In_2Te_3−InTe−Ge
alloys quenched in water from the melt region,
while Fig. 95 shows the isothermal segment at
400°C. As can be seen from Fig. 94, the phase
diagram is characterized by the phase regions α,
β, γ, ε, $\alpha + \beta$, $\alpha + \gamma$, $\alpha + \beta + \gamma$. Quenched alloys in the
GeTe−InGe segment containing 0-75 mol-% InTe
are monophasic solutions and have a lattice of
the NaCl type at an InTe content of 8-75 mol-%
and a lattice of the GeTe type at an InTe contant
below 8 mol-%. The solubilities of GeTe and
InTe are limited and amount to about 5 mol-%.
The isothermal segment at 400°C is characterized by the following regions: $\alpha + \beta$ (adjoin-
ing the GeTe side), $\alpha + \beta + \varepsilon$, $\beta + \delta + \varepsilon$, and $\beta +$
$\gamma + \delta$. No ternary compounds are found in the
system.

**Magnesium − Silicon − Germa-
nium**. The thermoelectric properties of the
pseudobinary Mg_2Si−Mg_2Ge system were determined by Labotz et al. [432]. X-ray and ther-
mal phase analyses established that Mg_2Si and Mg_2Ge exhibit complete mutual solubility. The
lattice constants and liquidus temperature were linear functions of composition. Measurements
of the electrical resistance and Hall effect showed that the current-carrier mobility at 300°K
was the same in a mixture of crystals as in the pure compounds, the maximum being somewhat
greater than 300 cm^2/V·sec. The forbidden-zone width decreases monotonically from $1.26 \cdot 10^{-19}$

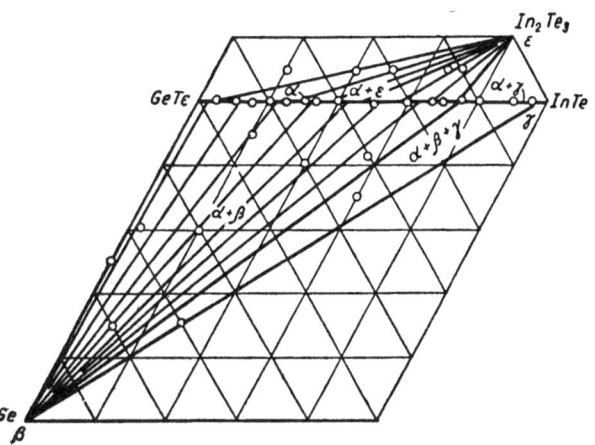

Fig. 94. Phase diagram of alloys in the GeTe−
In_2Te_3−InTe−Ge system [446].

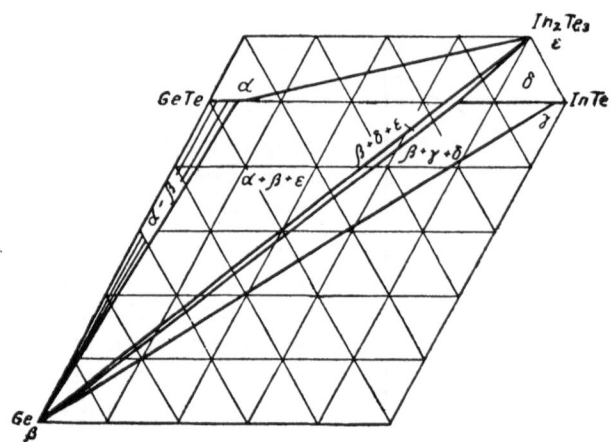

Fig. 95. Isothermal segment of the GeTe—In$_2$Te$_3$—
InTe—Ge system at 400°C [446].

J (0.78 eV) for Mg$_2$Si to 1.12·10^{-19} J (0.70 eV) for Mg$_2$Ge. Measurement of the thermal conductivity of the pseudobinary system showed that the coefficient of thermal conductivity at 300°K
was substantially lower for solid solutions than for the pure compounds. The coefficient for
Mg$_2$Si$_{0.4}$Ge$_{0.6}$ was 0.0268 W/cm·deg.

   Magnesium — Lead — Germanium. Differential thermal and x-ray analysis and
studies of the microstructure and microhardness of alloys in the Mg$_2$Pb—Mg$_2$Ge system [444]
were employed to construct a phase diagram for this system (Fig. 96). The diagram represents
a quasibinary segment of the ternary Mg—Pb—Ge system and shows that there is a peritectic
transformation at 555°C and that the components are partially soluble in one another (the solubility of Mg$_2$Ge in Mg$_2$Pb is about 2%, while that of Mg$_2$Pb in Mg$_2$Ge is about 7%).

   Magnesium — Tin — Germanium. Alloys in the Mg—Sn—Ge system have been
studied by the same methods as the Mg—Pb—Ge system [420, 462]. The data obtained were employed to construct a fusibility diagram for the Mg$_2$Sn—Mg$_2$Ge system (Fig. 97), which represents
a quasibinary segment of the ternary Mg—Sn—Ge system and shows that there is a peritectic

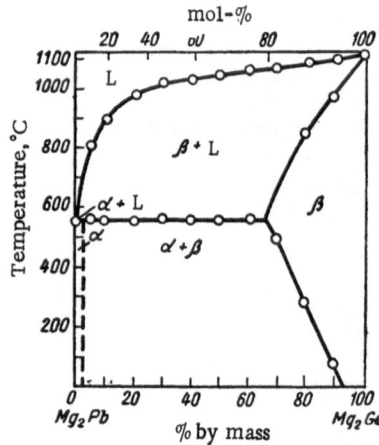

Fig. 96. Fusibility diagram of
the Mg$_2$Pb—Mg$_2$Ge system [444].

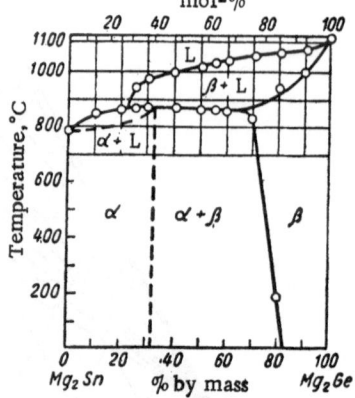

Fig. 97. Fusibility diagram of
the Mg$_2$Sn—Mg$_2$Ge system [449].

transformation at 870°C and that the solubilities of the components in one another are 33 and 18% by mass.

Makarov et al. [465] investigated the forbidden-zone width of alloy in the Mg_2Sn-Mg_2Ge and Mg_2Pb-Mg_2Ge systems; this factor governs the practical feasibility and applications of semiconductors. It was established that the forbidden-zone width (ΔE) for alloys in the Mg_2Sn-Mg_2Ge system ranges from 0.26 eV (100% Mg_2Sn) to 0.615 eV (100% Mg_2Ge), while that for alloys in the Mg_2Pb-Mg_2Ge system ranges from 0.08 eV (100% Mg_2Pb) to 0.615 eV (100% Mg_2Ge).

The variation in the electrical conductivity of alloys lying between the monophasic region and on the phase-region boundaries confirms the type of phase diagram previously established for these systems [444, 449].

M a n g a n e s e – S i l i c o n – G e r m a n i u m . Van Con [428] prepared solid solutions having the composition $Mn_5(Si_{1-x}Ge_x)_3$ and investigated their physical properties. The specimens were produced by fusion in a hermetically sealed quartz tube and purified by zone melting. The lattice constants of the solid solutions varied in the following manner, with x = 0, 0.2, 0.3, and 1:

	x = 0	x = 0.2	x = 0.3	x = 1
a, nm	0.691	0.694	0.697	0.717
c, nm	0.481	0.485	0.487	0.504

Investigation of the electrical resistance of specimens with the composition $Mn_5(Si_{1-x}Ge_x)_3$ established that their resistance drops monotonically as the temperature rises, the greatest decrease occurring over the temperature range 600-800°C. The Hall constant for Mn_5Si_3 is $1.4 \cdot 10^{-3}$ cm^3/C, while that for Mn_5Ge_3 and solid solutions lies between $1 \cdot 10^{-3}$ and $2 \cdot 10^{-3}$ cm^3/C at temperatures above the Curie point. The current-carrier mobility in a solid solution is 1 cm^3/V·sec, which does not permit us to draw precise conclusions regarding the conduction mechanism. The physical constants of specimens depend on the manner in which they are produced. Solid solutions, like the germanide Mn_5Ge_3, exhibit ferromagnetism. The authors conclude that substitution of Ge atoms for Si atoms does not alter the electronic structure of $Mn_5(Si_{1-x}Ge_x)_3$ solid solutions.

In working out the bonding pattern, the authors introduced two sublattices. The Mn_I atoms in the 4(d) positions form a linear chain of metallic bonds with one another, which is folded into sublattice·A. This confirms the metallic conductivity of these compounds at low temperatures.

The Mn_{II} and Ge atoms in the 6(g) positions form sublattice B, which forces us to introduce the orbital hybrids d^2s for Mn and sp^2 for Ge [430]. The magnetic moment of an Mn_{II} atom is 2 Bohr magnetons [431]. A $3d^3(d^2s)$ (p) configuration, leaving three electrons free in the d shell, is assumed to be possible.

C o p p e r – G a l l i u m – G e r m a n i u m . Metallographic and x-ray diffraction studies have been made of ternary copper-rich alloys containing up to 30 at-% Ga and 25 at-% Ge [358]. No ternary compounds have been found in the system.

M o l y b d e n u m (N i o b i u m) – T i n – G e r m a n i u m . Holleck et al. [359] conducted solubility studies in the Mo_3Ge-Nb_3Sn and Nb_3Ge-Nb_3Sn systems. They established that these systems contain continuous series of solid solution between the binary phases.

Figure 98 shows the variation in the lattice constants of alloys in these systems. Extrapolation of the curve for the solid solution Nb_3(Ge, Sn) to pure Nb_3Ge gives a lattice constant in good agreement with the data in the literature. No ternary compounds have been found in this system.

Table 13. Electrical Resistance, Temperature Coefficient of Thermal, and Termal conductivity of Ternary Ta(Nb)−Ge−Si germanides

Composition	Electrical resistance, μohm·cm (25°C)	Electrical resistance, μohm·cm (-196°C)	Temperature coefficient of thermal emf, μV/deg	Thermal conductivity, W/cm·deg (25°C)
$TaGe_{1,5}Si_{0,5}$	46	28	+11	—
$TaGeSi$	60	33	+14	0.21
$TaGe_{0,5}Si_{1,5}$	73	44	+17	0.24
$NbGe_{1,5}Si_{0,5}$	77	63	+17	0.19
$NbSiGe$	81	65	+22	0.16
$NbSi_{1,5}Ge_{0,5}$	60	47	+20	0.20

Niobium (Tantalum) − Silicon − Germanium. According to Brixner [360], continuous series of solid solutions between germanides and silicides of the AB_2 type are formed in these systems; their thermoelectric properties are shown in Table 13 [360].

Niobium − Aluminum − Tin − Germanium. Ageev and Shamrai [361] studied segments of the quaternary Nb−Al−Sn−Ge system corresponding to isoconcentrates with 83% and 75% niobium. All alloys containing 83% Nb are diphasic; one phase does not stain during etching and is a solid solution of aluminum, germanium, and tin in niobium (α-phase). The second phase is stained blue during etching and has been identified as having a β-W structure from x-ray diffraction data; it is a solid solution based on the compounds Nb_3Al, Nb_3Ge, and Nb_3Sn (the β-phase Nb_3[Al, Sn, Ge]). It was also established that only the β-phase is in equilibrium with the solid solution below 600°C and that the compounds Nb_3Al, Nb_3Ge, and Nb_3Sn form a continuous series of solid solutions.

The β-phase is the principal constituent of all alloys containing 75% Nb. Alloys other than the β-phase located close to the ternary Nb−Sn−Al diagram exhibit a niobium-based solid solution. Alloys with a higher germanium content are monophasic. More recently, Alekseevskii et al. [448] studied the critical temperature for transition of the β-phase to the superconductive state. Figure 99 shows the results of measurements for ternary alloys, while Fig. 100 shows the results of those for the quaternary system. It can be seen from these diagrams that the critical temperature varies uniformly with composition, falling as the germanium concentration increases. These authors believe that the principal parameter governing the critical temperature is the phase density at the Fermi surface, while the electron−phonon interaction parameter varies little.

Tin − Tellurium − Germanium. The SnTe−GeTe segment of the Sn−Te−Ge system has been studied by thermal, microstructural, and x-ray analysis [362].

It has been established that a continuous series of solid solutions is formed between the two nonisostructural compounds SnTe and GeTe, with a minimum in the fusibility diagram at a concentration of about 80% GeTe and a temperature of 700°C. The face-centered cubic lattice of the SnTe gradually passes into the face-centered rhombohedral lattice of the GeTe.

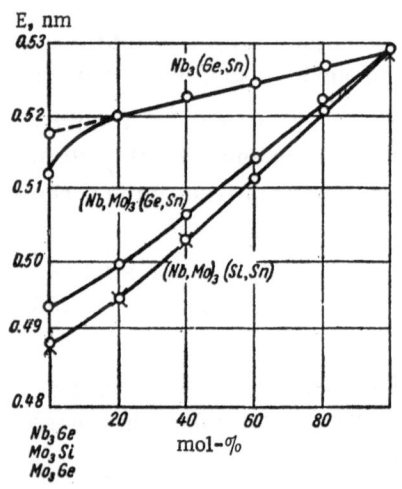

Fig. 98. Variation in lattice constants of alloys in the $Mo_3Ge − Nb_3Sn$ and $Nb_3Ge − Nb_3Sn$ system [359].

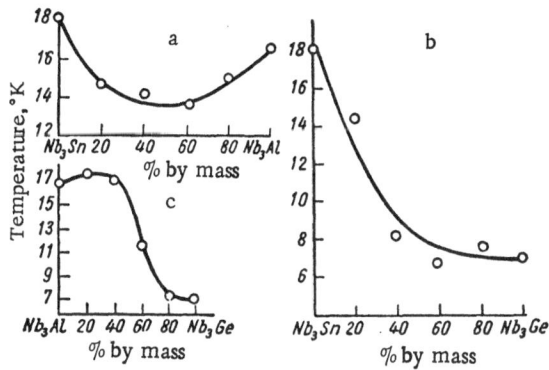

Fig. 99. Critical temperature for the following segments [448]; a) Nb_3Sn-Nb_3Al; b) Nb_3Sn-Nb_3Ge; c) Nb_3Al-Nb_3Ge.

Measurement of certain properties of the alloys showed that their microhardness and thermal emf vary regularly with alloy composition along the curve and reach maxima of about 600 MN/m^2 (60 kg/mm^2) at about 60% GeTe for the former and of about 33 $\mu V/deg$ at about 20% GeTe for the latter.

Conversely, the electrical conductivity decreases and reaches a minimum of about $0.2 \cdot 10^{-3}$ $ohm^{-1} \cdot cm^{-1}$ at about 60% GeTe. Figure 101 is a phase diagram of the SnTe-GeTe system.

Palladium (Platinum) - Carbon - Germanium. Stadelmaier and Hardy [363] prepared alloys in the ternary Pd-C-Ge system by fusing the high-purity components in graphite crucibles and studied them by microscopic and x-ray diffraction analysis.

In addition to the compound Pb_2Ge, these authors found a phase of unknown structure. No ternary compounds were detected in the system.

Alloys in the Pt-C-Ge system were prepared and studied in the same manner as those in the Pd-C-Ge system. The authors also failed to establish the existence of ternary compounds in this case.

Lead - Selenium - Germanium. The phase diagram of the PbSe-GeSe system (Fig. 102) was constructed from data obtained by thermal, microstructural, and x-ray analysis [429]. Since the phase GeSe is formed by a peritectic reaction, this system does not correspond to a quasibinary segment of the ternary Pb-Se-Ge system. The PbSe-GeSe segment overlaps two primary-crystallization fields: that for the solid solution not based on PbSe (the δ-phase) and that for $GeSe_2$. Primary crystallization of the δ-solid solution is complete at 630-617°C. A eutectic consisting of the δ-solid solution and $GeSe_2$ crystallizes below these temperatures.

The primary-crystallization region for $GeSe_2$ adjoins the triphasic region in which the peritectic reaction $L + GeSe_2 \rightleftharpoons \gamma$ occurs (phase field III). The γ-phase is a solid solution based on the high-temperature modification of GeSe. Crystallization of alloys in the central portion of the segment culminates in the invariant reaction $L + GeSe_2 \rightleftharpoons \delta + \gamma$, which takes place at 617°C.

The region occupied by solid solutions based on PbSe is somewhat less than 10 mol-% at 520°C. The solubility of PbSe in the high-temperature modification of GeSe reaches 20 mol-%.

The polymorphic-transformation temperature of GeSe decreases as the PbSe content rises. Eutectoid decay of the γ-solid solution and formation of an α-solid solution based on the low-temperature modification of GeSe and a δ-solid solution based on PbSe occur at about 460°C. The eutectoid contains almost 80 mol-% GeSe [429].

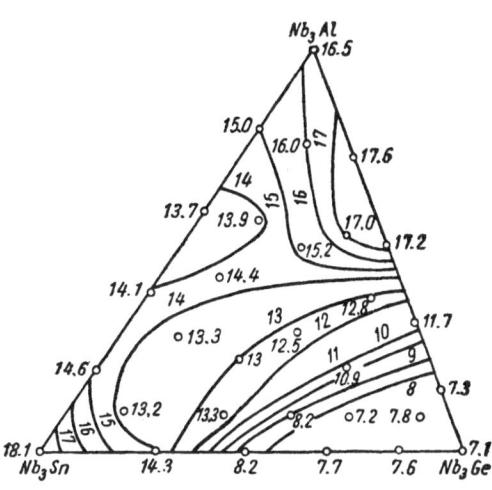

Fig. 100. Critical temperature (°K) for $Nb_3Sn-Nb_3Al-Nb_3Ge$ segment [448].

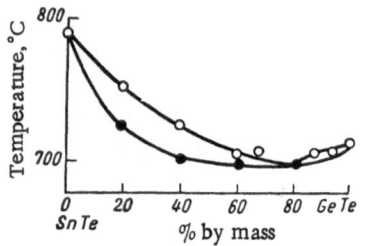

Fig. 101. Phase diagram of
the SnTe−GeTe system [362].

Selenium − Tellurium − Germanium. The
polythermal GeSe−GeTe segment was constructed in similar
fashion to the PbSe−GeSe segment (see p. 119). This segment
is not quasibinary and overlaps two primary-crystallization
fields: that of the $GeSe_xTe_{1-x}$ solid solution (the γ-phase) and
that of $GeSe_2$. Crystallization of the latter is complete at 670-
648°C. The peritectic reaction $L + GeSe_2 \rightleftharpoons GeSe_xTe_{1-x}$ takes
place at lower temperatures. Over the entire concentration
range, crystallization of the alloys culminates in formation of a
continuous series of solid solutions between the high-temperature modifications of GeSe and GeTe. The liquidus minimum corresponds to a GeTe content of about 40 mol-% and a tempera-
ture of 630°C. The phase transformations occurring in the alloys at lower temperatures were
studied by the dilatometric method. Thus, a phase transition corresponding to eutectoid decay,
$\gamma \rightarrow \alpha + \beta$, where α and β are solid solutions based on the low-temperature modifications of
GeSe and GeTe respectively, is observed at 370-380°C in alloys containing from 55 to 100 mol-%
GeTe. Three phase transitions occur in alloys with a GeTe content of less than 55 mol-%. The
first, at about 380°C, is related to the decay reaction $\gamma \rightarrow \alpha + \beta$, while the second is associated
with the eutectoid decay reaction $\gamma_1 \rightarrow \alpha + \gamma_2$ at temperatures of 460-470°C. The third transition, which takes place at higher temperatures, is associated with the shift from the diphasic
$\gamma_1 + \gamma_2$ region to the monophasic γ region.

The solubility of GeTe in GeSe at 425°C is less than 10 mol-%. Alloys containing between 10
and 100 mol-% GeTe have the polyhedral microstructure characteristic of the γ-solid solution [429].

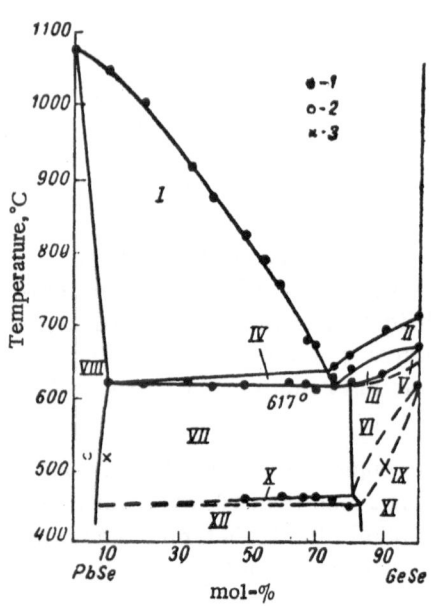

Fig. 102. The polythermal PbSe−GeSe
segment [429]. I) L+ δ = S.S.; II) L +
$GeSe_2$; III) L + $GeSe_2$ + γ = S.S.; IV) L + δ
+ $GeSe_2$; V) L + γ; VI) γ = S.S.; VII) δ +
γ; VIII) δ = S.S; IX) γ + α; X) γ + α + δ;
XI) α = S.S.; XII) δ + α; 1) thermal-
analysis data; 2) monophasic alloys; 3)
diphasic alloys.

Silver − Gold − Germanium. Ther-
mal and microscopic analysis of alloys* was employed to construct a diagram of the liquidus surface for the Ag−Au−Ge system [364] (Fig. 104).
It was established that the eutectic curve passes
through the concentration triangle from the Ag−
Ge side and terminates at the eutectic point of the
Au−Ge system. The primary-crystallization
phases in the Ag−Au−Ge system are the solid solution γ-Ag−Au and pure germanium.

Thorium − Silicon − Germanium.
Stecher et al. [365] investigated the Th−Si−Ge
system by x-ray diffraction analysis and melting-
point measurements. They established that continuous series of Th_3Si_2−Th_3Ge_2 and α-$ThSi_2$−
α-$ThGe_2$ solid solution are formed in this system,
their lattice constants increasing from a = 0.784
and c = 0.4152 nm to a = 0.796 and c = 0.4167 nm
and from a = 0.410 and c = 1.400 nm to a = 0.4141
and c = 1.422 nm respectively (values taken from
curves).

In the compound Th_2Ge, up to 60% of the germanium atoms are replaced by silicon atoms, the
lattice constants decreasing from a = 0.7412 and
c = 0.608 nm to a = 0.738 and c = 0.6025 nm.

*The alloys were prepared in the same manner as for
the Au−Sb−Ge system.

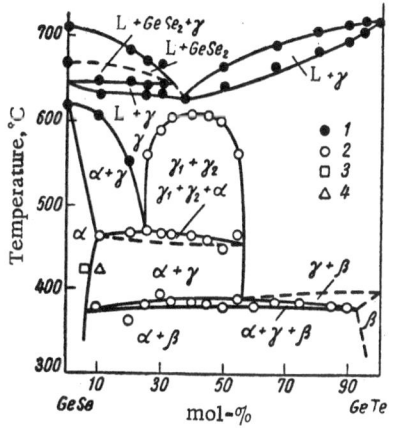

Fig. 103. Polythermal GeSe−GeTe segment [429]; 1) thermal analysis data; 2) dilatometric analysis data; 3) monophasic alloys; 4) diphasic alloys.

Up to 80 mol-% ThGe dissolves in ThSi, with an increase in the lattice constants of the latter. Only 12% of the germanium atoms in the ThGe are replaced by silicon atoms, with a decrease in lattice constants. Up to 45% of the germanium atoms in the compound $Th_{0.9}Ge_2$ are replaced by silicon atoms.

Figure 105 is a phase diagram of Th−Si−Ge system over the temperature range 900-1300°C.

Z i n c − A r s e n i c − I n d i u m − G e r m a n i u m. Giesecke and Pfister [366] studied the mutual solubility of the constituents of the $ZnGeAs_2$−InAs system.

It was established that about 80% $ZnGeAs_2$ dissolves in InAs, while about 20% InAs dissolves in $ZnGeAs_2$.

SYSTEMS IN WHICH CHEMICAL COMPOUNDS

ARE FORMED

N i t r o g e n − I r o n (C o b a l t, L i t h i u m) − G e r m a n i u m. Alloy specimens were prepared in the same manner as for the N−Mn−Ge system.

A ternary alloy with the composition $Co_{76}Ge_{19}N_5$ exhibited two face-centered cubic phases with the lattice constants a = 0.357 nm and a = 0.361 nm [343].

A ternary nitride with a tetragonal structure and the lattice constants a = 0.5318, c = 0.7733 nm, and c/a = 1.454 has been found in the N−Fe−Ge system [343].

According to Juza and Schulz [395], the ternary compound Li_5GeN_3, which is yellow in color and has the constant a = 0.9629 nm, exists in the N−Li−Ge system.

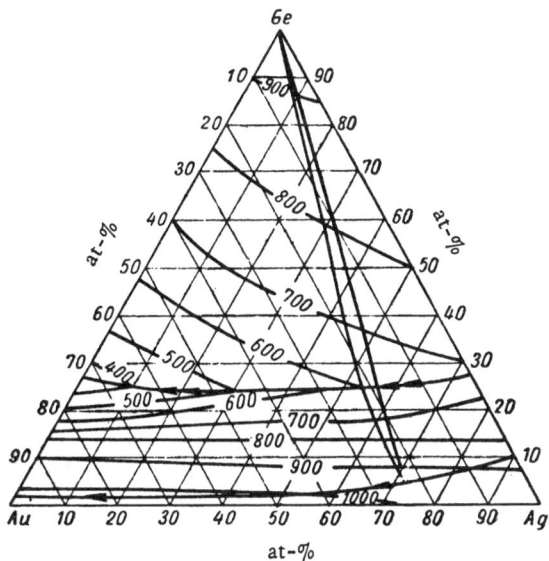

Fig. 104. Projection of the liquidus surface for the Ag−Au−Ge system [364].

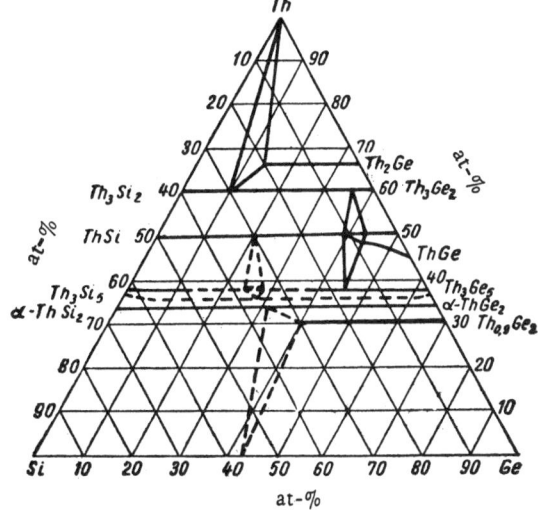

Fig. 105. Phase diagram of the Th−Si−Ge system [365].

Aluminum − Calcium − Germanium. According to Freundlich and Chartier [369], the Al−Ca−Ge system contains the two ternary compounds $CaAl_2Ge$ and $CaAl_2Ge_{1.5}$.

The compound $CaAl_2Ge_{1.5}$ has a cubic structure with lattice constant a = 0.715 mm. It is formed at 750°C as a result of the reaction between $CaAl_2$ and Ge.

Aluminum − Manganese − Germanium. Wernick et al. [370] studied the structural and magnetic relationships in the Al−Mn−Ge system and established the existence of the ternary compound MnAlGe. Thermal analysis of crystals of this compound showed that its melting point lies at 786°C, while solid-phase transformations occur at 584, 690, and 710°C.

The crystal structure of the compound was determined from powder patterns and high-resolution microphotography of monocrystals. The compound MnAlGe has a crystal lattice of the tetragonal type (D_{4h}-P4/nmm space group) with the constants a = 0.3910 and c = 0.5943 nm. The calculated density of the compound with two formal units in the elementary cell is in good agreement with the experimental value, amounting to 5.65 g/cm^3.

The MnAlGe phase is a magnetic compound with strong crystal anisotropy.

Aluminum − Palladium − Germanium. Various researchers [52, 371] have shown that the Al−Pd−Ge system contains the ternary compound Al_4GePd_5 with a cubic structure of the FeSi type and the lattice constant a = 0.4869 nm.

Barium − Selenium − Germanium. This system has been found to contain the ternary compound $BaGeSe_3$, which is formed as a result of the reaction between BaSe and $GeSe_2$ at temperatures of 600-1100°C [409]. X-ray analysis of monocrystals of this compound yielded the lattice constants (D_{2h}^{12} space group) a = 0.702, b = 1.413, and c = 0.605 nm. Its pycnometric density is 4.74 g/cm^3 and its x-ray density is 4.93 g/cm^3.

Beryllium − Copper − Germanium. According to x-ray data [487], cast alloys with the composition Be_2CuGe fused in evacuated quartz ampules contain a ternary Laves phase with a cubic structure of the $MgCu_2$ type, the composition Be_2CuGe, and the lattice constant a = 0.6018 nm.

Beryllium − Iron (Cobalt, Nickel) − Germanium. According to the data of Stadelmaier and Hofer [487], alloys with the composition Be_2MeGe in these systems contain elemental germanium and binary compounds in the Be−Fe (Co, Ni) system. However, a more recent investigation of the Be−Co−Ge system [488] discovered the ternary phase $BeCo_2Ge$ with a structure of the $MnCu_2Al$ type ($L2_1$) and the lattice constant a =0.546 nm. No phase of similar structure has been found in the Be−Ni−Ge system.

Vanadium (Niobium, Tantalum) − Nitrogen − Germanium. The V−Ge−N system has been found to contain a ternary H-phase (V_2GeN, with a structure of the Cr_2AlC type) and $V_5Ge_3N_X$, with a structure of the Mn_5Si_3 type ($D8_8$) [374].

The same investigation showed that the Ta−Ge−N system contains only the nitrogen-stabilized phase $Ta_5Ge_3N_X$, which has the same structure.

In a more recent investigation, Jeitschko et al. [375] determined the structure of compounds of the T_5M_3X type (where T is Nb or Ta, M is Ge, and X is N). The compounds $Nb_5Ge_3N_X$ and $Ta_5Ge_3N_X$ have a crystal structure of the $D8_8$ type with the lattice constants a = 0.760, c = 0.537 nm, and c/a = 0.707 and a = 0.7581, c = 0.515 nm, and c/a = 0.679 respectively.

Vanadium (Niobium, Tantalum) − Boron − Germanium. According to Rieger et al. [376], these systems contain an $MeGe_3B_X$ phase of the Mn_5Si_3 type, which is stabilized by boron and has the lattice constants a = 0.7310, c = 0.4978 nm, and c/a = 0.681 for $V_5Ge_3B_X$, a = 0.7710, c = 0.5330 nm, and c/a = 0.6912 for $Nb_5Ge_3B_X$, and a = 0.7658, c = 0.5300 nm, and c/a = 0.6921 for $Ta_5Ge_3B_X$.

Vanadium – Iron – Germanium. This system has been found to contain the compound VFe_2Ge with a cubic structure of the $MnCu_2Al$ type and the lattice constant $a = 0.5774$ nm.*

A ternary compound with a structure of the α-Mn type (x-phase) and the lattice constant $a = 0.895$ nm has been detected in specimens with the composition $V_{30}Fe_{55}Ge_{15}$ annealed at 800°C (Smith et al [454]).

Vanadium – Cobalt – Germanium. According to Smith et al. [454], this system contains three new compounds in addition to the previously known compound VCoGe [144]. The first has a cubic structure of the α-Mn type with the lattice constant $a = 0.8940$ nm, the second has the composition $V_{34}Co_{39}Ge_{27}$ and a teteragonal structure (being isostructural with the δ-phase in the V–Fe–Si system), and the third has the composition $V_{25}Co_{70}Ge_5$.

Vanadium – Nickel – Germanium. According to Gladyshevskii et al. [192], the V–Ni–Ge system contains the compound $V_3Ni_5Ge_2$ with a structure or superstructure of the α-Mn type and the lattice constant $a = 0.892$ nm.

Smith et al. [454] established that $V_3Ni_5Ge_2$ is decomposed when the temperature is reduced to 800°C. Lines corresponding to a second ternary compound are clearly visible in the x-ray patterns of alloys annealed at 800°C. The positions and relative intensities of these lines and the lattice constant $a = 0.6517$ nm indicate that the compound in question has a structure of the β-M type (π-phase). The V–Ni–Ge system therefore contains ternary compounds with structures of the α-Mn and β-Mn types, the former existing at high temperatures and the latter at low temperatures.

Vanadium – Iron (Cobalt, Nickel) – Nitrogen – Germanium. According to Smith et al. [454], ternary compounds with a structure of the β-Mn type (π-phases) are formed in these systems, having lattice constants a of 0.6512, 0.6496, and 0.6511 nm respectively.

Vanadium (Niobium, Tantalum, Chromium, Molybdenum) – Carbon – Germanium. Perri [377], who investigated alloys in the Cr–Ge–C system, observed a carbon-containing phase with the composition $Cr_5Ge_3C_X$ in one specimen; its powder pattern indicated a tetragonal cell.

A more recent investigation [372] confirmed and clarified the composition of this phase. It was found to be in equilibrium with a phase of the $D8_8$ type and germanium and must be assigned the formula Cr_2GeC (H-phase).

The lattice constants of the Cr_2GeC phase are $a = 0.2954$, $c = 1.208$ nm, and $c/a = 4.091$. The least interatomic distances are Cr–C = 0.199, Cr–Ge = 0.263 nm, Cr–Cr = 0.269 nm, and Ge–Ge = 0.295 nm. The compound has an x-ray density of 6.86 g/cm^3.

A compound of simlar composition was obtained in the V–Ge–C system. The lattice constants of the V_2GeC phase are $a = 0.300$, $c = 1.225$ nm and $c/a = 4.083$. Its x-ray density is 5.76 g/cm^3.

The shortest interatomic distances, calculated from these parameters, are V–C = 0.203 nm, V–Ge = 0.265 nm, V–V = 0.273 nm, and Ge–Ge = 0.300 nm.

The Ta–Ge–C system has been found to contain a carbon-stabilized phase with the composition $Ta_5Ge_3C_X$ and the lattice constants $a = 0.7586$, $c = 0.5217$ nm, and $c/a = 0.687$. The carbon content of this phase reaches 5-10 at-%. No H-phase has been found in the system.

According to Jeitschko et al. [375], phases with same composition also exist in the Nb-Ge–C and Mo–Ge–C systems. The compounds $Nb_5Ge_3C_X$ and $Mo_5Ge_3C_X$ have a structure of the

---

*V. Ya. Markiv, Author's abstract of dissertation [in Russian], L'vov (1966).

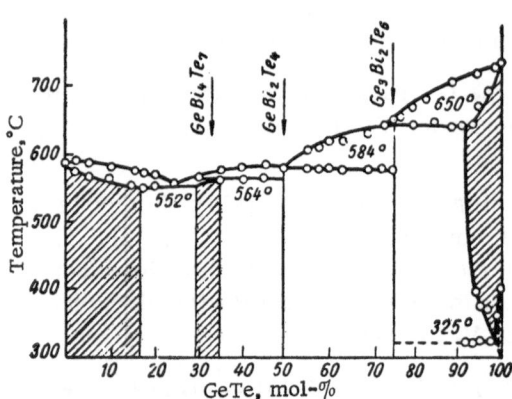

Fig. 106. Phase diagram of the Bi_2Te_3– GeTe system [379].

Mn_5Si_3 type with the lattice constants a = 0.7628, c = 0.539 nm, and c/a = 0.707 and a = 0.7369, b = 0.5047 nm, and c/a = 0.685 respectively.

Vanadium–Tantalum–Germanium.
According to Gangeberger et al. [433], this system contains a ternary compound with the approximate composition Ta_2V_3Ge; it is a typical Laves phase (C14 type) with the lattice constants a = 0.5048, c = 0.8278 nm, and c/a = 1.640.

Bismuth–Tellurium–Germanium.
Abrikosov and Danilova-Dobryakova [379] employed microscopic, thermal, and x-ray analysis of alloys prepared by joint fusion of other alloys in evacuated quartz ampules to study the ternary Bi–Te–Ge system. The initial alloys were previously prepared Bi_2Te_3 and GeTe. After their microstructure was examined, the final alloys were annealed in ampules filled with argon to a pressure of about 49.03 kN/m^2 (0.5 atm), with a holding time of 1500 h at 500°C for alloys containing up to 50 mol-% GeTe and 550°C for alloys containing 50–100 mol-% GeTe. The data obtained were employed to construct a phase diagram for the Bi_2Te_3–GeTe system (Fig. 106). As can be seen from this diagram, there is a solid-solution region extending from the Bi_2Te_3 side to 17.4 mol-% GeTe at 500°C. The solubility of Bi_2Te_3 in GeTe was not determined. Three ternary compounds were found in the system: Bi_4Te_7Ge, Bi_2Te_4Ge, and $Bi_2Te_6Ge_3$. They are formed by peritectic reactions at 564, 584, and 650°C respectively.

Table 14. Properties of Ternary Compounds

Compound	Composition, mol-%		Microhardness		Forbidden-zone width	
	Bi_2Te_3	GeTe	MN/m^2	kg/mm^2	J	eV
Bi_4Te_7Ge . . .	66.7	33.3	300	30	$0.336 \cdot 10^{-19}$	0.21
Bi_2Te_4Ge . . .	50.0	50.0	470	47	$0.368 \cdot 10^{-19}$	0.23
$Bi_2Te_6Ge_3$ . . .	25.0	75.0	1190	119	—	—

Table 14 shows certain properties of the ternary compounds.

Tungsten–Silicon–Germanium.
Nowotny et al. [380] detected a ternary chemical compound with the approximate composition $W_{0.4}Si_{0.35}Ge_{0.25}$ in this system. Up to 15 at-% of the silicon in the WSi_2 phase is replaced by germanium, while up to 5 at-% of that in the W_5Si_3 phase (T1) is replaced. Figure 107 is a phase diagram of the ternary W–Si–Ge system.

Gallium–Iron (Cobalt, Nickel)–Germanium.
Esslinger and Schubert [52] studied the Ga–Fe–Ge, Ga–Co–Ge, and Ca–Ni–Ge systems and established the existence of the compounds Fe_2GaGe (with a lattice of the B-35 type), Co_4GaGe_3 (with a structure of the B-20 type and the lattice constant a = 0.4639 nm), and $Ni_{50}Ga_{42}Ge_8$ (with a structure of the B-20 type and the lattice constant a = 0.4649 nm).

Cadmium–Copper–Germanium.
According to Teslyuk and Markiv [382], this system contains the ternary phase $CdCu_{1.5}Ge_{0.5}$ (λ_2) with the lattice constant a = 0.7115 nm.

Cadmium–Arsenic–Germanium.
This system has been found to contain the compound $CdGeAs_2$ [384–386], which exhibits semiconductive properties. Goryunova et al. [387,

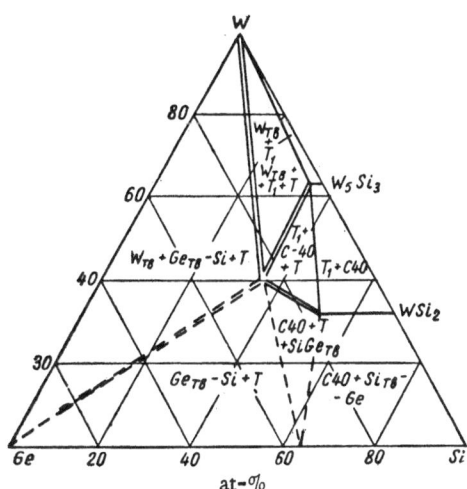

Fig. 107. Phase diagram of the W—Si—Ge system [380].

388] produced n- and p-type simple crystals of this compound and studied certain of its properties, establishing that it has a chalcopyrite structure with the lattice constants a = 0.5942, c = 1.1217 nm, and c/a = 1.887. Its melting point is 665°C, its microhardness is 4710 MN/m^2 (471 kg/mm^2), and its forbidden-zone width at 300°K, as determined by optical measurements, is 0.84·10^{-19} J (0.53 eV). The Hall hole and electron mobilities in specimens with a current-carrier concentration n = 10^{17} cm^{-3} are 20-25 and 800-1000 cm^2/V·sec at room temperature. The thermal emf of an electronic specimen at 300°K is 190 µV/deg.

Cobalt — Boron — Germanium. According to Fruchart and Fruchart [389], the Co—B—Ge system contains the ternary compound $Co_{21}Ge_2B_6$, which is isomorphic with $Cr_{23}C_6$ and has a cubic structure with the lattice constant a = 1.0499 nm.

This compound was produced by heating a mixture of cobalt, germanium, and boron powders in evacuated quartz ampules for several days at 750°C or for 8 h at 850°C. It is ferromagnetic and has a Curie point at 238°C. The magnetic moment of each cobalt atom is 0.55 µV. The authors believe that the existence of an $Me_{23}X_6$ compound and a compound of the cementite type cannot be attributed solely to the atomic-radius ratio of the metal and nonmetal; it is also associated with the number of valence electrons in the nonmetal and the number of 3d electrons in the transition metal. The nonmetal is the electron donor in this case.

Cobalt (Manganese, Nickel) — Iron — Germanium. Isomorphic FeMeGe compounds with a structure of the NiAs type are formed in these systems [136, 137]. The lattice constants are given in Table 20.

Cobalt (Nickel) — Manganese — Germanium. The Ni—Mn—Ge and Co—Mn—Ge systems have been found to contain two ternary compounds of the MeMnGe type [136, 137, 390] with the lattice constants a = 0.4066, c = 0.5391 nm, and c/a = 1.326 and a = 0.4042, c = 0.5251 nm, and c/a = 1.299 respectively, as well as a compound of the Me_2MnGe type [390] with a Cu_2MnAl structure.

The compound Co_2MnGe is a homogeneous ternary compound with the lattice constant a = 0.5744 nm. It has a broad homogeneity region (20-33.3 at-% Mn, 55-41.7 at-% Co, and 25 at-% Ge) and is stable at temperatures of 300, 400, 600, and 800°C. At 800°C, $MnCo_2Ge$ is in equilibrium with the solid solution based on α-Co and with the compound Mn_2Co_3Ge (the λ_1-phase), which has a structure of the $MgZn_2$ type and the lattice constants a = 0.4803, c = 0.7739 nm, and c/a = 1.611. When the temperature is reduced to 500°C, Mn_2Co_3Ge decomposes into two phases: a solid solution based on α-Co and an H-phase.

Busev [300] believes that the composition of the compound Ni_2MnGe differs somewhat from this formula and that it has the lattice constant a = 0.5701 nm.

The homogeneity region of this compound has not been established. In addition to the compounds mentioned above, x-ray diffraction and microscopic studies of the Ni—Mn—Ge system enabled various researchers [287, 382, 392] to establish the existence of the compound $Mn_6Ni_{16}Ge_7$, which has a structure of the $Mg_6Cu_{16}Si_7$ type and the lattice constant a = 1.141 nm, and of two compounds with a Laves-phase structure: $MnNi_{1.25}Ge_{0.75}$ (λ_1) with the lattice constants a = 0.4856, c = 0.7635 nm, and c/a = 1.572 and $MnNi_{1.5}Ge_{0.5}$ (λ_2) with the lattice constant a = 0.6762 nm.

It has been reported [393] that $MnCo_2Ge$ and Ni_2MnGe have magnetic properties (the Curie temperature for $MnCo_2Ge$ is 557°C).

Lithium − Copper − Germanium. According to Teslyuk and Oleksiv [394], the Li−Cu−Ge system contains a ternary Laves phase with the approximate composition $LiCu_{1.5}Ge_{0.5}$, a structure of the $MgCu_2$ type (O_h^7-Fd3m space group), and the lattice constant a = 0.706 nm.

Schuster [391] reports the existence of the compound $LiCu_2Ge$, which has a structure of the feldspar type and the lattice constant a = 0.5892 nm. It is a reddish-violet substance with a lustrous metallic surface and readily dissolves in concentrated oxidizing acids.

Lithium (Zinc) − Arsenic − Germanium. According to Juza and Schulz [395], the Li−As−Ge system contains the chemical compound Li_5GeAs_3 with the lattice constant a = 0.6092 nm. This compound was produced by heating the mixed components in quartz ampules at 700-850°C for 2-3 h. It is black in color and has an x-ray density of 3.27 g/cm^3.

The Zn−As−Ge system has been found to contain a chemical compound of the $MeAs_2Ge$ type with a chalcopyrite structure and the lattice constants a = 0.567, c = 1.1152 nm, and c/a = 1.967 [384].

Magnesium − Nickel − Germanium. According to x-ray and microscopic data [192, 382], this system contains the compounds $MgNi_{1.6}Ge_{0.4}$ (λ_2), which has a structure of the Laves type and the lattice constant a = 0.6911 nm, and $Mg_6Ni_{16}Ge_7$, which has a structure of the $Mg_6Cu_{16}Si_7$ type and the lattice constant a = 1.1532 nm. The latter compound was prepared by direct fusion of the 99.9%-pure components in corundum crucibles under a KCl−LiCl flux, with subsequent annealing at 400-600°C for 1 month.

Manganese (Magnesium) − Copper − Germanium. According to x-ray diffraction data [192], the Mn−Cu−Ge system contains the compound $MnCu_{1.5}Ge_{0.5}$, which has a structure of the $MgZn_2$ type and the lattice constants a = 0.4929, c = 0.7864 nm, and c/a = 1.595.

The ternary Mg−Cu−Ge system was investigated in order to determine whether it contains ternary compounds [192, 381].

Alloys having the compositions $MgCu_{1.5}Ge_{0.5}$, $Mg_6Cu_{16}Ge_7$, MgCuGe, and $MgCu_4Ge$ were prepared and hardened at 600°C. Microstructural and x-ray diffraction studies showed that only the alloy with the composition $MgCu_{1.5}Ge_{0.5}$ was homogeneous. This compound has a hexagonal structure of the $MgZn_2$ type (D_{6h}^4-P6_3/mmc space group) and the lattice constants a = 0.5043, c = 0.8008 nm, and c/a = 1.588 or a = 0.5033-0.5087, c = 0.8007-0.8119, and c/a = 1.588-1.596 for compositions in the range $MgCu_{1.55-1.25}Ge_{0.15-0.75}$ [192]. The elementary cell contains 12 atoms; the Mg atoms are in the 4(f) positions with z = 1/16, while the Cu and Ge atoms are in the 2(a) and 6(h) positions with x = 5/6.

Manganese (Iron, Cobalt, Nickel) − Carbon − Germanium. According to Huetter and Stadelmeier [396], these systems contain phases with a cubic structure of the $CaTiO_3$ type; their lattice constants are a = 0.3867-0.3877 nm for CMn_3Ge, a = 0.3657 nm for $C_{0.45}Fe_3Ge$, a = 0.3597-0.3627 nm for $C_{0.25}Co_3Ge$, and a = 0.3577 nm for C_xNi_3Ge.

Jeitschko et al. [372] investigated specimens with the composition Mn_2GeC and found no H-phases. In addition to pure germanium, pure carbon, and the compound Mn_5Ge_3, specimens with this composition contained a small amount of an unidentified phase.

No H-phase was detected in the Fe−Ge−C system. However, an $Fe_{1+x}Ge$ phase of the NiAs type, which apparently contained carbon, appeared in addition to α-Fe and $FeGe_2$. The lattice constants of this compound were a = 0.4032, c = 0.5024 nm, and c/a = 1.246, as against a = 0.4017 and c = 0.5005 nm for $Fe_{1.7}Ge$ [397].

Copper — Gold — Germanium. King et al. [398] employed x-ray diffraction analysis to study the variation in the axial ratio (c/a) of the densely packed face-centered lattice of the ternary ξ-phase of the Cu−Ge−Au system as a function of alloy electron concentration at 475 and 400°C. They established that up to 20 at-% of the Cu in this system can be replaced by gold at constant Gc contents between 12.7 and 13.5 at-%. The substitution reduces the value of c/a for all electron concentrations, the decrease with rising copper (or gold) content becoming larger as the electron concentration increases. The authors attribute the decrease in c/a to formation of a Brillouin phase at high electron concentrations.

Copper — Nickel — Germanium. The Cu−Ni− Ge system was studied by microscopic and x-ray analysis of alloys prepared by fusion in evacuated quartz ampules [397], followed by annealing at 500°C.

The system was found to contain the ternary compounds $Ni_3Cu_2Ge_2$ (with a hexagonal lattice), $Ni_{15}Cu_{65}Ge_{20}$ (with a lattice of the β-Mn type and the constant a =0.6272 nm), and $Ni_{15}Cu_{58}Ge_{22}$ (with a lattice of the γ-brass type).

Copper — Sulfur (Selenium, Tellurium) — Germanium. Palatnik et al. [400] employed x-ray analysis and study of certain physical properties (density, microhardness, and melting point) to investigate alloys in the ternary system Cu−R−Ge (where R is S, Se, or Te) and found compounds of the A_2BC_3 type (where A is Cu, B is Ge, and C is S, Se, or Te).

All the specimens were synthesized in evacuated quartz ampules and subjected to zone purification, which eliminated any deviations from stoichiometric composition.

The compounds detected have a lattice of the diamond type and belong to the class of daltonides with a congruent melting point. Table 15 gives the lattice constants and certain physical properties of these compounds [400, 437].

Table 15. Lattice Parameters and Certain Physical Properties of Ternary Compounds

Compound	Lattice constant, nm	Density, g/cm³		Microhardness, $H_{\mu_{100}}$		T_m, °C
		Calculated	Experimental	MN/m²	kg/mm²	
Cu_2S_3Ge	0.530	4.39	4.36	6070 3900*	607 390*	948 956*
Cu_2Se_3Ge . . .	0.555	5.65	5.65	5320 3070*	532 307*	770 783*
Cu_2Te_3Ge . . .	0.595	6.11	6.14	4730	473	504

*Balanevskaya et al. [437].

It can be seen from this table that the physical properties depend monotonically on the lattice constant a, decreasing uniformly as the latter increases.

According to Palatnik et al. [400], these compounds have semiconductive properties; the same authors made a separate study of the properties in question [401] and established that the compounds are covalent, having a lattice of the diamond type with tetrahedral coordination. All the compounds have p-type conductivity. Table 16 shows the properties of these germanides.

Berger and Balanevskaya [402] measured the thermal expansion α, thermal conductivity χ, and modulus of elasticity E of the semiconductive ternary compounds Cu_2S_3Ge and Cu_2Se_3Ge. Table 17 gives the results of these measurements.

It has been shown [473] that the compound Cu_2Se_3Ge is capable of forming solid solutions with GaAs, these containing up to 40 mol-% Cu_2Se_3Ge.

Table 16. Physical Properties of Ternary Compounds

Compound	Hall constant, cm^3/C	Main current-carrier mobility, $cm^2/V \cdot sec$	Main current-carrier conc., $1/cm^3$	Conductivity, $ohm^{-1} \cdot cm^{-1}$	Coeff. of thermal emf, $\alpha, \mu V/deg$	Forbidden-zone width ΔE	
						J	eV
Cu_2S_3Ge ..	—	360*	3.0*	1.9 / 17.3*	100—300 / 439*	$0.48 \cdot 10^{-19}$ / $1.76 \cdot 10^{-19}$	0.3 / 1.1*
Cu_2Se_3Ge .	44.0	1870 / 438*	$1.7 \cdot 10^{17}$ / 1.5*	50 / 5.71*	70—100 / 143*		— / 0.8*
Cu_2Te_3Ge .	—	—	—	$1.4 \cdot 10^3$	10	$1.28 \cdot 10^{-19}$	—

*Balanevskaya et al. [437].

Table 17. Thermal Expansion, Modulus of Elasticity, and
Thermal Conductivity of Ternary Compounds

Compound	$\alpha \cdot 10^6$, deg^{-1}	ρ, g/cm^3	$E \cdot 10^{-12}$		$\chi \cdot 10^{-3}$, $W/cm \cdot deg$
			N/m^2	dN/m^2	
Cu_2S_3Ge ..	7.2	4.36	0.14	1.40	7.70
Cu_2Se_3Ge ..	8.4	5.65	0.091	0.91	6.91

Copper − Zinc (Mercury) − Selenium − Germanium. According to Hahn and Schulze [409], these systems contain the quaternary compounds $Cu_2ZnGeSe_4$ and $Cu_2HgGeSe_4$, which have a structure of the stannite type (D_{2d}^{1}-I42m space group) with the lattice constants a = 0.562, c = 1.106 nm, and c/a = 1.967 and a = 0.5694, c = 1.102 nm, and c/a = 1.937 respectively. Their pycnometric densities are 5.48 and 6.52 g/cm^3, while their x-ray densities are 5.52 and 6.66 g/cm^3.

Molybdenum − (Iron, Cobalt, Nickel) − Germanium [192]. These systems have been found to contain compounds with an $MgZn_2$ structure: $Mo(Fe, Ge)_2$ with the lattice constants a = 0.483, c = 0.786 nm, c/a = 1.63, $Mo(Co, Ge)_2$ with the lattice constants a = 0.480, c = 0.773 nm, and c/a = 1.61, and $Mo(Ni, Ge)_2$ with the lattice constants a = 0.482, c = 0.770 nm, and c/a = 1.60.

Borusevich [403], who employed x-ray diffraction analysis, confirmed the existence of compounds with an $MgZn_2$ structure in these systems, their compositions and lattice constants differing slightly from those given above.

According to Skolozdra et al. [176], these systems contain ternary compounds with the following compositions (in at-%): $Mo_{50}Fe_{33}Ge_{17}$ (a = 1.115, c = 2.006 nm, and c/a = 1.80), $Mo_{50}Co_{30}Ge_{20}$ (a = 1.114, c = 1.999 nm, and c/a = 1.79), and $Mo_{45}Ni_{32}Ge_{23}$ (a = 1.115, c = 2.011 nm, and c/a = 1.81).

Molybdenum − Carbon − Germanium. Jeitschko et al. [372, 373] studied the Mo−Ge−C system in order to determine whether it contains ternary compounds. Only a $D8_8$ phase was detected. X-ray studies of alloys containing 52 at-% Mo, 22 at-% Ge, and 26 at-% C showed that ternary cystals producing a relatively complex pattern were present in addition to a small amount of Mo_2Ge_3. The authors in question [372] assume that this ternary phase must be richer in the transition metal than an H-phase.

Arsenic − Tellurium − Germanium. Vitreous compounds (alloys) with semiconductive properties are formed in this system. According to Panus et al. [474, 475], the current-carrier sign in vitreous alloys with the compositions $AsTeGe_x$ (where x = 0-0.4) and

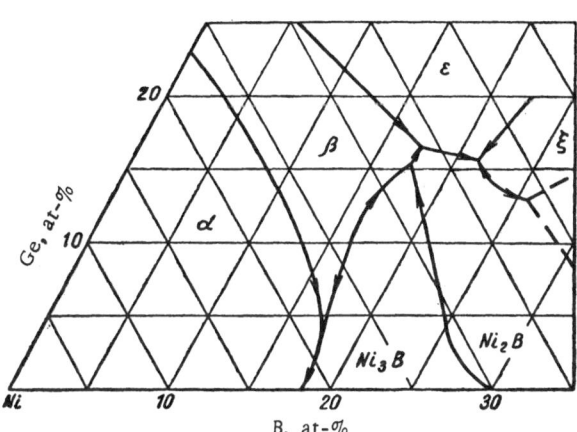

Fig. 108. Projection of the liquidus surface of the Ni−Ge−B system [410].

AsTe$_y$Ge$_{0.3}$ (where y = 0.8 or 1.5) corresponds to n-type conductivity and the Hall current-carrier mobility is 0.043-0.065 cm^2/V·sec at 20°, increasing exponentially with temperature; the activation energy of mobility ranges from 0.22 to 0.12 eV and decreases as the structure of the glass becomes more ideal and as its crystallization capacity decreases. The coefficient of thermal emf at 20° ranges from 1030 to 2000 μV/deg.

Niobium (Tantalum) − Iron (Cobalt, Nickel) − Germanium. Teslyuk et al. [383] made x-ray and microstructural analyses of ternary alloys with the composition R$_2$X$_3$Ge, where R is Nb or Ta and X is Fe, Co, or Ni. The alloys were prepared by fusing weighed portions of the high-purity (99.9% basic component) metals in aluminum oxide crucibles in an argon atmosphere in a Tamman furnace. They were then heat-treated in evacuated quartz ampules at 600°C for two weeks.

All the alloys hardened at 600°C were ternary Laves phases with a structure of the MgZn$_2$ (λ_1) type. Thus, in the Nb(Ta)−Fe−Ge systems, these phases were ternary solid solutions of germanium in the binary compounds NbFe$_2$ and TaFe$_2$. In the Nb(Ta)−Co−Ge systems, the ternary λ_1-phases were individual ternary compounds, as well as solid solutions based on the corresponding high-temperature modifications of the binary Laves phases in the Nb−Co and Ta−Co systems. The ternary λ_1-phases detected in the Nb(Ta)−Ni−Ge systems were new ternary Laves phases of the MgZn$_2$ type.

Table 20 [383] gives the lattice constants for the observed ternary λ-phases with the composition RX$_{1.5}$Ge$_{0.5}$.

Niobium − Chromium − Germanium. Gangeberger et al. [433] established that this system contains a ternary compound of the Laves-phase type with the composition NbCrGe and the lattice constants a = 0.4949 and c = 8.095 nm.

Nickel − Boron − Germanium. Stadelmaier and Lee [410] employed microscopic and x-ray analysis to investigate alloys in the Ni−Ge−B system containing up to 25 at-% Ge and up to 35 at-% B. The data obtained were used to construct a diagram of the liquidus surface (Fig. 108) and an isothermal section of the system at 800°C (Fig. 109). These authors established that, in addition to the previously known phases based on the binary nickel−germanium and nickel−boron system (α, β, ε, Ni$_3$B, and Ni$_2$B), the ternary Ni−Ge−B system contains the ternary phase Ni$_{60}$Ge$_{15}$B$_{20}$ (ξ-phase), whose structure has not been determined, and the unstable ternary phase Ni$_{75}$Ge$_5$B$_{20}$ with structure of the Cr$_{23}$C$_6$ type.

R − Nickel − Germanium and R − Cobalt − Germanium. Gladyshevskii et al. [406] conducted x-ray diffraction studies of a number of ternary alloys prepared by fusion of the pure metals in aluminum oxide crucibles in an argon atmosphere in order to determine whether ternary germanides with structures of the Mg$_6$Cu$_{16}$Si$_7$ type (O$_h^5$−Fm3m space group) were present. All the alloys were annealed at 800°C for 120 h to even out their compositions.

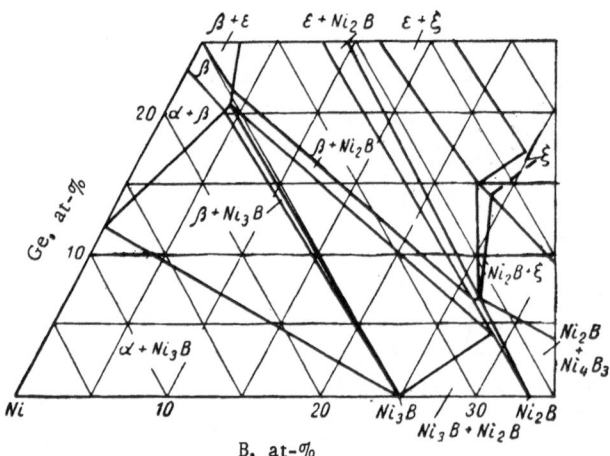

Fig. 109. Isothermal section of the nickel corner of
the Ni−Ge−B system at 800°C [410].

It was established that the R−Ni−Ge (where R is Sc, Ti, Nb, Hf, Ta, or Zr) and R−Co−Ge (where R is Nb, Hf, Ta, or Zr) systems contain compounds with this type of structure (see Table 20).

The $Mg_6Cu_{16}Si_7$ structure belongs to the class of structures with coordination numbers and is a superstructure of the Th_6Mn_{23} type. The $Mg_6Cu_{16}Si_7$ elementary cell contains 116 atoms. The Mg atoms occupy positions with a coordination number of 17, the Si atoms occupy positions with coordination numbers of 14 and 12, and the Cu atoms occupy positions with coordination numbers of 13 and 12.

The Ti−Ni−Ge, Zr−Co−Ge, and Hf−Co−Ge systems contain compounds with a tetragonal structure of the $TiNiSi_2$ type ($14/mmm-D_{4h}^{17}$ space group): $TiNiGe_2$ (a = 1.304 and c = 0.498 nm), $ZrCoGe_2$ (a = 1.320 and c = 0.523 nm), and $HfCoGe_2$ (a = 1.316 and c = 0.521 nm).

Lead − Sulfur − Germanium. Elli and Mugnoli [407] employed thermal, microscopic, and x-ray analysis to investigate the $PbS−GeS_2$ system at GeS_2 contents between 0 and 55.89 mol-%, using specimens produced by fusion of the components. They established the existance of two chemical compounds: $PbGeS_3$ and Pb_2GeS_4.

The compound $PbGeS_3$ is a yellow substance that melts at 593°C, while Pb_2GeS_4 forms bright red crystals that melt at 621°C.

Specimens with a content of more than 56 mol-% GeS_2 are unstable and decompose when heated, liberating GeS_2.

Silver − Selenium (Tellurium) − Germanium. According to Balanevskaya et al. [437], these systems contain the ternary compounds Ag_2Se_3Ge and Ag_2Te_3Ge, whose physical and semiconductive properties are shown in Table 18.

Antimony − Tellurium − Germanium. Abrikosov and Danilova-Dobryakova [378] investigated this system by microscopic, thermal, and x-ray analysis of alloys prepared by joint fusion of GeTe and Sb_2Te_3 in quartz ampules evacuated to 13.33 mN/m² (10^{-4} mm Hg).

The microscopic and thermal data were used to construct a phase diagram for the $Sb_2Te_3−$ GeTe system (Fig. 110). As can be seen from this diagram, a solid-solution region extending

Table 18. Physical and Semiconductive Properties of Ternary Compounds

Compound	Density, g/cm³	Melting point, °C	Electrical conductivity at 300 °K, ohm⁻¹·cm⁻¹	Current-carrier concentration, cm⁻³·10⁻¹⁷	Current-carrier mobility, cm²/V·sec	Forbidden-zone width			
						From electrical measurements		From optical measurements	
						J	eV	J	eV
Ag₂Se₃Ge ...	4.66	540	27	2	850	1.44·10⁻¹⁹	0.9	1.45·10⁻¹⁹	0.91
Ag₂Te₃Ge ...	5.12	330	92	8	720	0.4·10⁻¹⁹	0.25	—	—

to about 8 mol-% GeTe at 300°C is formed on the basis of Sb_2Te_3. The solubility increases when the temperature is raised, reaching about 13 mol-% at 593°C. The solid-solution region on the GeTe side extends to 10 mol-% Sb_2Te_3 at 550°C. The solubility of Sb_2Te_3 in GeTe decreases as the temperature is raised, amounting to about 5 mol-% Sb_2Te_3 at 400°C. The solubility of Sb_2Te_3 in the low-temperature modification of GeTe is very small. In addition, the system has been found to contain the three ternary compounds Sb_4Te_7Ge, Sb_2Te_4Ge, and $Sb_2Te_5Ge_2$. All three compounds are formed by peritectic reactions, at temperatures of 605, 615, and 630°C respectively.

Figures 111 and 112 show the variation in the electrical conductivity and thermal emf of annealed alloys as a function of composition. There is a sharp increase in conductivity and a decrease in thermal emf in the solid-solution region based on Sb_2Te_3. The authors attribute the increase in conductivity to the change in current-carrier concentration resulting from the difference in the valence of Sb and Ge, assuming that some of the germanium in the Sb_2Te_3-based solid solution in replaced by antimony. The opposite holds in the GeTe-solid solution: electrical conductivity decreases and thermal emf increases. Table 19 shows some of the physical properties of the ternary compounds at room temperature.

Titanium — Cobalt — Germanium. Gladyshevskii et al. [194] studied this system, using alloys prepared by fusion of the high-purity components in an inert atomosphere in an

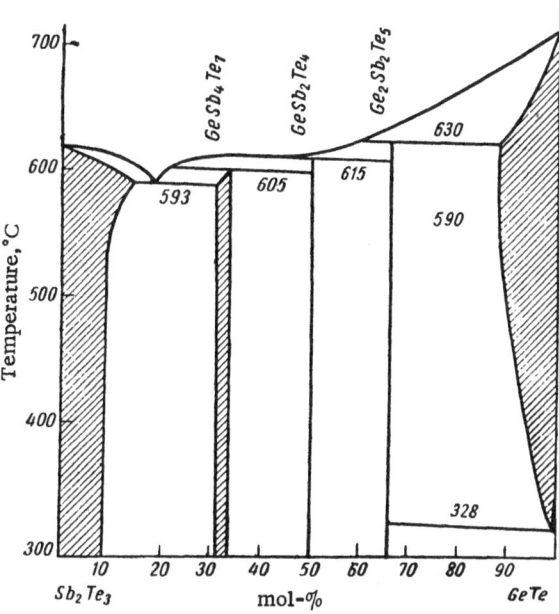

Fig. 110. Phase diagram of the Sb_2Te_3—GeTe system [378].

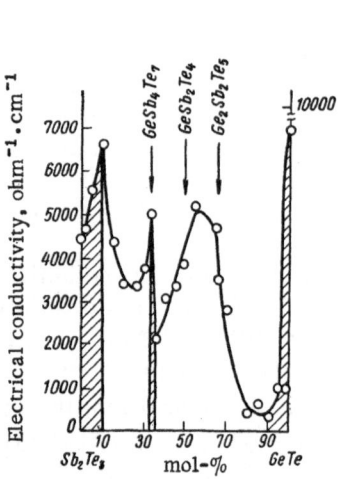
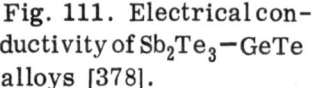

Fig. 111. Electrical conductivity of Sb_2Te_3–GeTe alloys [378].

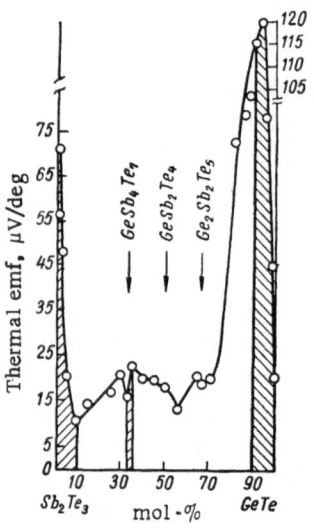

Fig. 112. Thermal emf of Sb_2Te_3–GeTe alloys [378].

electric furnace. The purpose of this investigation was to detect chemical compounds and determine their structure.

X-ray and microstructural analysis established that the Ti–C–Ge system contains the compound $TiCo_2Ge$, which has a structure of the $MnCu_2Al$ type (O_h^5-Fm3m space group) and the lattice constant a = 0.5823 nm.

In addition to the aforementioned compound, Markiv et al. [408] established the existence of the compound Ti_2Co_3Ge, which has a structure of the $MgZn_2$ type and the lattice constants a = 0.4867 and c = 0.7590 nm.

Titanium (Zirconium) – Gallium – Germanium. According to Raman and Schubert [367], the Ti–Ga–Ge system contains one chemical compound with the approximate composition Te_2GaGe_3, having a structure of the $ZrSi_2$ type (C49) and the lattice constants a = 0.369, b = 1.409, and c = 0.367 nm.

The Zr–Ga–Ge system has been found to contain the compounds $Zr_4Ga_{11}Ge$ and $ZrGa_{0.1}Ge_{0.9}$

The compound $Zr_4Ga_{11}Ge$ has the lattice constants a = 0.369, c = 0.910 nm and c/2a = 1.17.

Titanium – Carbon – Germanium. Jeitschko et al. [372] stated that an H-phase probably exists. They subsequently produced this compound and calculated its lattice constants [373]. The Ti_2GeC was obtained by hot pressing and subsequent annealing at 1300°C for 12 h in

Table 19. Physical Properties of Ternary Compounds

| Compound | Composition, mol-% | | T_m, °C | Thermal emf α, μV/deg | Electrical conductivity σ, $ohm^{-1} \cdot cm^{-1}$ | Microhardness | |
	Sb_2Te_3	GeTe				$\frac{MN}{m^2}$	kg/mm^2
Sb_4Te_7Ge . . .	66.7	33.3	605	27	5000	450	45
Sb_2Te_4Ge . . .	50	50	615	18.6	3800	568	56.8
$Sb_2Te_5Ge_2$ . . .	33.3	66.7	630	28.5	3500	650	65

Table 20. Structure of Ternary Compounds

System	Germanide	Structure	Structure type	Lattice constants, nm				Density, g/cm³		Reference
				a	b	c	c/a	Pycnometric	X-ray	
Al–Ca–Ge	$CaAl_2Ge$	—	—	—	—	—	—	—	—	[369]
	$CaAl_2Ge_{1.5}$	Cubic	—	0.715	—	—	—	—	—	[369]
Al–Mn–Ge	$AlMnGe$	Tetragonal	—	0.3910	—	0.5943	1.520	—	5.650	[370]
Al–Pd–Ge	Al_4Pd_5Ge	Cubic	$FeSi$	0.4869	—	—	—	—	—	[52, 371]
Ba–Se–Ge	$BaSe_3Ge$	Rhombic	—	0.702	1.413	0.605	—	4.74	4.98	[409]
Be–Co–Ge	$BeCo_2Ge$	Cubic	$MnCu_2Al$	0.546	—	—	—	—	—	[488]
Be–Cu–Ge	Be_2CuGe	Cubic	$MgCu_2$	0.6018	—	—	—	—	—	[487]
Cd–As–Ge	$CdAs_2Ge$	Tetragonal	$CuFeS_2$	0.5943	—	1.1217	1.887	—	—	[384–387]
Cd–Cu–Ge	$CdCu_{1.5}Ge_{0.5}$ (λ_2)	Cubic	$MgCu_2$	0.7115	—	—	—	—	—	[382]
Ce–Ni–Ge	$CeNi_2Ge_2$	Tetragonal	$CeGa_2Al_2$	0.415	—	0.984	2.370	—	—	[447]
Co–B–Ge	$Co_{21}B_6Ge_2$	Cubic	$Cr_{23}C_6$	1.0499	—	—	—	—	—	[389]
Co–C–Ge	$Co_3C_{0.25}Ge$	»	$CaTiO_3$	0.3597–0.3627	—	—	—	—	—	[396]
Co–Fe–Ge	$CoFeGe$	Hexagonal	$MnGeNi$	0.3978	—	0.5028	1.264	—	—	[136, 137]
Co–Mn–Ge	$CoMnGe$	»	$MnGeNi$	0.4042	—	0.5251	1.299	—	—	[136, 137, 390]
	Co_2Mn_2Ge	Cubic	Cu_2MnAl	0.5744	—	—	—	—	—	[136, 137, 390]
	Co_3Mn_2Ge	Hexagonal	$MgZn_2$	0.4803	—	0.7789	1.611	—	—	[391]
Cr–C–Ge	Cr_2CGe	—	Cr_2AlC'	0.2954	—	1.208	4.091	—	6.860	[372]
Cu–Ni–Ge	$Cu_2Ni_3Ge_2$	Hexagonal	Ni_5As_5	0.6863–0.6974	—	1.2525–1.2645	1.823–1.815	6.52	6.66	[399]
Cu–Hg–Se–Ge	Cu_2HgSe_4Ge	—	—	0.5694	—	1.102	1.937	—	—	[409]
Cu–Ni–Ge	$Cu_{65}Ni_{15}Ge_{20}$	Cubic	β-Mn	0.6272	—	—	—	—	—	[399]
	$Cu_{58}Ni_{15}Ge_{22}$	»	γ-Cu_5Zn_8	—	—	—	—	—	—	[399]
Cu–S–Ge	Cu_2S_3Ge	»	—	0.530	—	—	—	—	—	[400]
Cu–Se–Ge	Cu_2Se_3Ge	»	—	0.555	—	—	—	—	—	[400]
Cu–Te–Ge	Cu_2Te_3Ge	»	—	0.595	—	—	—	—	—	[400]
Cu–Zn–Se–Ge	Cu_2ZnSe_4Ge	—	—	0.5622	—	1.106	1.967	—	—	[409]
Eu–Ni–Ge	$EuNi_2Ge_2$	Tetragonal	$CeGa_2Al_2$	0.414	—	1.010	2.44	—	—	[447]
Fe–C–Ge	$Fe_3C_{0.45}Ge$	Cubic	$CaTiO_3$	0.3657	—	—	—	—	—	[396]
Ga–Fe–Ge	$GaFe_2Ge$	Hexagonal	$CoSn$	0.501	—	0.4043	0.807	—	—	[52]
Ga–Co–Ge	$GaCo_4Ge_3$	Cubic	$FeSi$	0.4639	—	—	—	—	—	[52]
Ga–Ni–Ge	$Ga_{42}Ni_{50}Ge_8$	»	$FeSi$	0.4649	—	—	—	—	—	[52]
Hf–Co–Ge	$Hf_6Co_{16}Ge_7$	»	$Mg_6Cu_{16}Si_7$	1.1566	—	—	—	—	—	[406]
	$HfCoGe_2$	Tetragonal	$TiNiSi$	1.316	—	0.521	—	—	—	*
Hf–Ni–Ge	$Hf_6Ni_{16}Ge_7$	Cubic	$Mg_6Cu_{16}Si_7$	1.1606	—	—	—	—	—	[406]
Li–As–Ge	Li_3As_3Ge	»	CaF_2	0.6092	—	—	—	—	3.270	[395]

Table 20 (Continued)

System	Germanide	Structure	Structure type	a	b	c	c/a	Pycnometric	X-ray	Reference
Li—Cu—Ge	LiCu₁.₅Ge₀.₅ (λ₂)	Cubic	MgCu₂	0.706	—	—	—	—	—	[394]
	LiCu₂Ge			0.5892	—	—	—	—	—	[391]
Mg—Cu—Ge	MgCu₁.₅Ge₀.₅ (λ₁)	Hexagonal	MgZn₂	0.5043	—	0.8008	1.588	—	—	[192, 381]
Mg—Ni—Ge	MgNi₁.₅Ge₀.₄ (λ₂)	Cubic	MgCu₂	0.6911	—	—	—	—	—	[192, 382]
	Mg₆Ni₁₆Ge₇	"	Mg₆Cu₁₆Si₇	1.1532	—	—	—	—	—	[192, 382]
Mn—C—Ge	Mn₃CGe	Cubic	CaTiO₃	0.3867–0.3877	—	—	—	—	—	[396]
Mn—Cu—Ge	MnCu₁.₅Ge₀.₅ (λ₁)	Hexagonal	MgZn₂	0.4929	—	0.7864	1.595	—	—	[192]
Mn—Fe—Ge	MnFeGe	"	MnGeNi	0.4104	—	0.5223	1.273	—	—	[136, 137]
Mo—C—Ge	Mo₅CₓGe₃	"	Mn₅Si₃	0.7369	—	0.5047	0.685	—	—	[375]
Mo—Fe—Ge	Mo(Fe, Ge)₂	"	MgZn₂	0.4830	—	0.7860	1.630	—	—	[192, 403]
	~Mo₅Fe₃Ge₂	—	—	1.115	—	2.006	1.80	—	—	[176]
Mo—Co—Ge	Mo(Co, Ge)₂	Hexagonal	MgZn₂	0.480	—	0.7730	1.610	—	—	[192]
	Mo₅Co₃Ge₂	—	—	1.114	—	1.999	1.79	—	—	[176]
Mo—Ni—Ge	Mo(Ni, Ge)₂	Hexagonal	MgZn₂	0.4820	—	0.770	1.600	—	—	[192]
	~Mo₈Ni₃Ge₂	—	—	1.115	—	2.011	1.81	—	—	[176]
N—Co—Ge	N₅Co₇₆Ge₁₀	Cubic	—	0.357	—	—	—	—	—	[343]
N—Fe—Ge	—	"	—	0.361	—	—	—	—	—	[343]
	—	Tetragonal	—	0.5318	—	0.7733	1.454	—	—	[343]
N—Li—Ge	N₃Li₅Ge	Cubic	N₂(Li₃Al)	0.9629	—	—	—	—	—	[395]
Nb—B—Ge	Nb₅BₓGe₃	Hexagonal	Mn₅Si₃	0.7710	—	0.5330	0.691	—	—	[376]
Nb—C—Ge	Nb₅CₓGe₃	"	Mn₅Si₃	0.7628	—	0.5390	0.707	—	—	[375]
Nb—Co—Ge	Nb₆Co₁₆Ge₇	Cubic	Mg₆Cu₁₆Si₇	1.1495	—	—	—	—	—	[406]
	NbCo₁.₅Ge₀.₅ (λ₁)	Hexagonal	MgZn₂	0.4860	—	0.7832	1.612	—	—	[383]
Nb—Ni—Ge	Nb₆Ni₁₆Ge₇	Cubic	Mg₆Cu₁₆Si₇	1.1425	—	—	—	—	—	[406]
	NbNi₁.₅Ge₀.₅ (λ₁)	Hexagonal	MgZn₂	0.4875	—	0.7879	1.696	—	—	[383]
Nb—N—Ge	Nb₅NₓGe₃	"	Mn₅Si₃	0.760	—	0.5370	0.707	—	—	[375]
Ni—B—Ge	Ni₆₀B₂₀Ge₁₅	—	—	—	—	—	—	—	—	[410]
	Ni₇₅B₃₀Ge₅	Cubic	Cr₂₃C₆	—	—	—	—	—	—	[410]
Ni—C—Ge	Ni₃CₓGe	Cubic	CaTiO₃	0.3577	—	—	—	—	—	[396]
Ni—Fe—Ge	NiFeGe	Hexagonal	MnGeNi	0.4016	—	0.5082	1.266	—	—	[136, 137]
Ni—Mn—Ge	NiMnGe	"	MnGeNi	0.4066	—	0.5391	1.326	—	—	[136, 137, 390]
	Ni₂MnGe	Cubic	Cu₄MnAl	0.5701	—	—	—	—	—	[136, 137, 390]
	Ni₁₆Mn₆Ge₇	"	Mg₆Cu₁₆Si₇	1.141	—	—	—	—	—	[287, 382, 392]

Table 20 (Continued)

System	Germanide	Structure	Structure type	a	b	c	c/a	Density, g/cm³ Pycnometric	Density, g/cm³ X-ray	Reference
	$Ni_{1.25}MnGe_{0.75}(\lambda_1)$	Hexagonal	$MgZn_2$	0.4856	—	0.7635	1.572	—	—	[287, 382, 392]
	$Ni_{1.4}MnGe_{0.5}(\lambda_2)$	Cubic	$MgCu_2$	0.6762	—	—	—	—	—	[287, 382, 392]
P—Cd—Ge	$CdGeP_2$	Tetragonal	$CuFeS_2$	0.5738	—	1.035	1.878	—	—	[384, 385]
P—Cu—Ge	P_3CuGe_2	Cubic	ZnS	0.5375	—	—	—	—	—	[384]
P—Mg—Ge	P_2MgGe	Cubic	ZnS	0.5652	—	—	—	—	—	[384]
P—Zn—Ge	P_2ZnGe	Tetragonal	$CuFeS_2$	0.546	—	1.075	1.970	—	—	[384]
P—Li—Ge	P_3Li_5Ge	Cubic	CaF_2	0.5891	—	—	—	—	2.180	[395]
Rh—Ga—Ge	Rh_5GaGe_4	Cubic	$FeSi$	0.4831	—	—	—	—	—	[52]
Sc—Ni—Ge	$Sc_6Ni_{16}Ge_7$	Cubic	$Mg_6Cu_{16}Si_7$	1.160	—	—	—	—	—	[406]
Ta—B—Ge	$Ta_5B_xGe_3$	Hexagonal	Mn_5Si_3	0.7658	—	0.530	0.692	—	—	[376]
Ta—C—Ge	$Ta_5C_xGe_3$	Hexagonal	Mn_5Si_3	0.7586	—	0.5217	0.687	—	—	[372]
Ta—Co—Ge	$Ta_6Co_{16}Ge_7$	Cubic	$Mg_6Cu_{16}Si_7$	1.1420	—	—	—	—	—	[406]
	$TaCo_{1.5}Ge_{0.5}(\lambda_1)$	Hexagonal	$MgZn_2$	0.4875	—	0.7817	1.603	—	—	[383]
Ta—Ni—Ge	$Ta_6Ni_{16}Ge_7$	Cubic	$Mg_6Cu_{16}Si_7$	1.1446	—	—	—	—	—	[406]
	$TaNi_{1.5}Ge_{0.5}(\lambda_1)$	Hexagonal	$MgZn_2$	0.4845	—	0.7836	1.617	—	—	[383]
Ta—N—Ge	$Ta_5N_xGe_3$	Hexagonal	Mn_5Si_3	0.7581	—	0.5150	0.679	—	—	[374, 375]
Ti—C—Ge	Ti_2CGe	Hexagonal	—	0.3079	—	1.2930	4.199	—	5.640	[373]
Ti—Co—Ge	$TiCo_2Ge$	Cubic	$MnCu_2Al$	0.5823	—	—	—	—	—	[194]
	$Ti_2Co_3Ge(\lambda_1)$	Hexagonal	$MgZn_2$	0.4867	—	0.7590	1.560	—	—	[408]
Ti—Ga—Ge	Ti_2GaGe_3	Rhombic	$ZrSi_2$	0.3690	1.409	0.3670	—	—	—	[367]
Ti—Ni—Ge	$Ti_6Ni_{16}Ge_7$	Cubic	$Mg_6Cu_{16}Si_7$	1.1470	—	—	—	—	—	[406, 194]
	$TiNiGe_2$	Tetragonal	$TiNiSi_2$	1.304	—	0.498	—	—	—	*
Th—Co—Ge	$ThCo_2Ge_2$	Tetragonal	$ThCo_2Si_2$	0.4109	—	0.9934	—	—	—	[330]
Th—Cr—Ge	$ThCr_2Ge_2$	Tetragonal	$ThCr_2Si_2$	0.411	—	1.083	—	—	—	[330]
Th—Cu—Ge	$ThCu_2Ge_2$	Tetragonal	$ThCu_2Si_2$	0.4152	—	1.0140	—	—	—	[330]
Th—Fe—Ge	$ThFe_2Ge_2$	Tetragonal	$ThFe_2Si_2$	0.4098	—	1.0222	—	—	—	[330]
Th—Mn—Ge	$ThMn_2Ge_2$	Tetragonal	$ThMn_2Si_2$	0.4084	—	1.0930	—	—	—	[330]
Th—Ni—Ge	$ThNi_2Ge_2$	Tetragonal	$ThNi_2Si_2$	0.4161	—	0.9677	—	—	—	[330]
V—B—Ge	$V_5B_xGe_3$	Hexagonal	Mn_5Si_3	0.7310	—	0.4978	0.681	—	—	[376]
V—C—Ge	V_2CGe	—	Cr_2AlC	0.3001	—	1.225	4.083	—	—	[372]
V—Fe—Ge	VFe_2Ge	Cubic	$MnCu_2Al$	0.5774	—	—	—	—	5.760	*
V—N—Ge	V_2NGe	—	Cr_2AlC	—	—	—	—	—	—	[374]
V—Ni—Ge	$V_3Ni_6Ge_2$	Cubic	$\alpha—Mn$	0.892	—	—	—	—	—	[192]
Zn—As—Ge	$ZnAs_2Ge$	Tetragonal	$CuFeS_2$	0.567	—	1.1152	1.967	—	—	[384]
Zn—Co—Ge	$ZnCo_2Ge$	Cubic	$MnCu_2Al$	0.574	—	—	—	—	—	[488]
Zn—Ni—Ge	$ZnNi_2Ge$	Cubic	$MnCu_2Al$	0.544	—	—	—	—	—	[489]
Zr—Ca—Ge	Zr_4Ca_6Ge	—	—	0.389	—	0.910	2.34	—	—	[367]
	$ZrCa_{2.1}Ge_{0.9}$	—	—	—	—	—	—	—	—	[367]
Zr—Co—Ge	$Zr_6Co_{16}Ge_7$	Cubic	$Mg_6Cu_{16}Si_7$	1.1628	—	—	—	—	—	[406]
	$ZrCoGe_2$	Tetragonal	$TiNiSi_2$	1.320	—	0.523	—	—	—	*
Zr—Fe—Ge	$Zr_2Fe_3Ge(\lambda_1)$	Hexagonal	$MgZn_2$	0.5020	—	0.8169	1.627	—	—	[408]
Zr—Mo—Ge	$ZrMo_{1.5}Ge_{0.5}(\lambda_1)$	Hexagonal	$MgZn_2$	0.540	—	0.875	1.620	—	—	[409]

* V. Ya. Markiv, Dissertation, L'vov, 1966.

a purified-argon atmosphere. The alloys were homogeneous and were identified as an H-phase having the lattice constants a = 0.3079, c = 1.293 nm, and c/a = 4.199. Its x-ray density was 5.64 g/cm^3.

In addition to Ti$_2$GeC, Wolfsgruber et al. [480] established the existence of the compound Ti$_3$GeC$_2$, which was produced by hot pressing and subsequent high-vacuum annealing at 1200°C for 20 h. This compound has a hexagonal structure (isostructural with Ti$_3$SiC$_2$) and the lattice constants a = 0.3077, c = 1.776 nm, and c/a = 5.770.

Thorium — Transition Metal — Germanium. Systems of the Th−R−Ge type (where R is Cr, Mn, Fe, Co, Ni, or Cu) have been found to contain compounds with the composition ThMe$_2$Ge$_2$ which are isostructural with ThMe$_2$Si$_2$ [330]. They have a tetragonal structure (J4/mmm space group) with two formal units in the elementary cell. Their lattice constants are given in Table 20.

Cerium (Europium) — Nickel — Germanium. Bodak et al. [447] employed x-ray diffraction analysis to study these systems and establish the existence of the compounds CeNi$_2$Ge$_2$ and EuNi$_2$Ge$_2$, which have a structure of the CeGa$_2$Al$_2$ or BaAl$_4$ type and the lattice constants a = 0.415, c = 0.984 nm, and c/a = 2.37 and a = 0.414, c = 1.010 nm, and c/a = 2.44 respectively.

Haschke et al. [463] studied the Ce−Si−Ge system.

Zinc — Cobalt (Nickel) — Germanium. According to Parthe [488], these systems contain the compounds ZnCo$_2$Ge and ZnNi$_2$Ge, which have a structure of the MnCu$_2$Al type and the lattice constants a = 0.574 nm and a = 0.544 nm respectively.

Zirconium — Iron — Germanium. According to Markiv et al. [408], the Zr−Fe−Ge system contains the compound Zr$_2$Fe$_3$Ge, which has a structure of the MgZn$_2$ type and the lattice constants a = 0.5020 and c = 0.8169 nm.

Zirconium — Molybdenum — Germanium. The Zr−Mo−Ge system contains a ternary Laves phase with the composition ZrMo$_{1.5}$Ge$_{0.5}$, a structure of the MgZn$_2$ type, and the lattice constants a = 0.540, c = 0.875 nm, and c/a = 1.62.

Phosphorus — Lithium — Germanium. According to Juza and Schulz [395], the P−Li−Ge system contains the chemical compound Li$_5$GeP$_3$ with the lattice constant a = 0.5891 nm.

This compound was produced by heating the mixed components at 700-1000°C for 2-3 h in quartz ampules and is a cinnamon-brown "valence" compound with an x-ray density of 2.18 g/cm^3.

The P−Cu−Ge, P−Mg−Ge, P−Cd−Ge, and P−Zn−Ge systems have been described by Folberth and Pfister [384] (see Table 20).

CHAPTER 7

CHEMICAL PROPERTIES OF GERMANIDES

METAL GERMANIDES

Germanides of Group-I Metals. Alkali-metal germanides, like the corresponding silicides, are extremely susceptible to the action of moisture. Thus, both lithium−germanium compounds react with moist air at room temperature to form Li_2Co_3; Li_6Ge_2 is less active in this respect than Li_4Ge [59].

The potassium germanides KGe and KGe_4 react with air and water, but less vigorously than potassium silicides.

Rhubidium and cesium germanides are stable only in a dry inert atmosphere. They react with air and water, decomposing into the corresponding alkali hydroxide and an amorphous brownish mass, apparently $(GeH)_x$, which is oxidized to germanium dioxide (GeO_2) by nitric acid [3, 49].

Copper Germanides. Addition of germanium increases the corrosion resistance of copper. Thus, alloys containing up to 25% Ge are soluble only in concentrated nitric acid, while alloys with a higher germanium content are soluble only in aqua regia [299].

Germanides of Group-II Metals. A fresh fracture surface in an alloy with the composition Mg_2Ge has a fine-grained structure and is silvery-gray in color. This compound decomposes in water; after 5 min in air, a fracture suface turns yellow and gives off an odor of germanium hydride (GeH_2).

Calcium germanides, like magnesium germanides, are chemically unstable.

The compound CaGe has a silvery-white color in an inert medium. In air it decomposes into an orange-red powder and $Ca(OH)_2$ or $CaCO_3$ [101].

The compound Ca_2Ge also reacts readily with air, an orange powder being formed on its surface [102].

Germanides of Group-IV Metals.* The germanides of the titanium subgroup are rather stable compounds. Table 21 presents the results of a study of the solubility of titanium, zirconium, and hafnium germanides with the composition Me_5Ge_3 on boiling for 1 h.

It can be seen from these data that germanides of the titanium subgroup decompose only in a mixture of nitric and hydrofluoric acids. They dissolve to a considerable extent in hydrochloric acid (1:1 and concentrated), sulfuric acid (concentrated), and aqua regia.

Germanides of Group-V Metals. Carpenter and Searcy [113], who studied the chemical properties of alloys in the niobium-germanium system, found that the chemical com-

---

*The chemical properties of certain germanides have been studied by O. I. Popova.

Table 21. Solubility of Certain Germanides in Different Aggressive Media on Boiling for 1 h, %

Compound	HCl (1.19)	HCl (1 : 1)	H_2SO_4 (1.84)	H_2SO_4 (1 : 4)	HNO_3 (1.4)	HNO_3 (1 : 1)	HNO_3 + HCl
Ti_5Ge_3	81	80	82	63	*	*	99
Zr_5Ge_3	83	85	88	46	—	21	91
Hf_5Ge_3	79	85	90	34	16	21	91
Ta_5Ge_3	3	2	30	0	—	—	1
Mo_3Ge_2	3	2	—	26	—	100	100
Mo_2Ge_3	3	3	—	—	—	100	100
$MoGe_2$	4	0	—	—	—	100	100

*Solution of most of the germanium, accompanied by hydrolysis.

Compound	HCl + H_2O_2	H_2O_2	Bromine water	HNO_3 + HF	H_2SO_4 + $(NH_4)S_2O_8$	NH_4OH + H_2O_2	H_2SO_4 + K_2SO_4
Ti_5Ge_3	83	100	5	100	88	—	—
Zr_5Ge_3	87	—	—	100	—	—	—
Hf_5Ge_3	83	0,5	0	100	27	—	—
Ta_5Ge_3	1	3	—	100	—	—	100
Mo_3Ge_2	80	100	81	100	—	75	—
Mo_2Ge_3	88	99,5	79	100	—	61	—
$MoGe_2$	5	—	71	100	—	—	—

pounds Nb_3Ge and $NbGe_2$ are unstable in molten soda, in 30% hydrogen peroxide, and in cold sodium hydroxide solutions. Neither compound reacts with hydrochloric acid or aqua regia.

The tantalum germanide Ta_5Ge_3 in more stable with respect to ordinary chemical reagents. It is insoluble in hydrochloric and sulfuric acids and aqua regia and partially soluble in mixtures of molten sodium or ammonium fluoride with hydrogen peroxide. It is completely soluble in mixtures of concentrated sulfuric acid with potassium sulfate (wet fusion) and of hydrofluoric acid with nitric acid (Table 21).

Germanides of Group-VI Metals. Searcy et al. [118] made a detailed study of the chemical properties of germanium-molybdenum compounds. They established that all the phases react very slowly with nonoxidizing acids and alkalis. They are less resistant to oxidizing agents and high-temperature oxidation than the corresponding silicides.

Thus, concentrated or dilute sulfuric, hydrochloric, and hydrofluoric acids in cold or hot form react only slightly with these compounds. Oxidizing agents, however, readily attack them. For example, 30% hydrogen peroxide or cold nitric acid in any concentration completely dissolves Mo_3Ge and Mo_3Ge_2, the solution taking on a yellow color. The compounds Mo_2Ge_3 and $MoGe_2$ do not dissolve completely, producing white sediments (probably GeO_2). A molten mixture of nitrates and carbonates decomposes these compounds at an almost explosive rate.

Germanides of Group-VII Metals. The compound $ReGe_2$ is stable in aggressive media, reacting only with hot concentrated sulfuric acid and molten sodium hydroxide [222].

Germanides of Group-VIII Metals. An alloy with the composition $FeGe_2$ corrodes in air. The corrosion resistance of other alloys in air increases with their germanium content [88, 89].

Lanthanide Germanides. The technique for chemical analysis of lanthanum germanides reduces to the following [147]. A weighed portion of the germanide (0.4-0.5 g) is dissolved in dilute (1:4) nitric acid. After complete solution has been obtained, the lanthanum and germanium in aliquot portions are determined. The lanthanum content is found by the complexometric method, i. e., titration with trylon B, using arsenazo as the indicator [412]. The germanium is determined after binding of the lanthanum into a soluble complex with trylon B, using a volumetric procedure, i.e., titration with alkali in the presence of mannitol, with phenolphthalein as the indicator [413].

According to Lyutaya and Goncharuk [147], lanthanum digermanide ($LaGe_2$) readily reacts with water, while $LaGe$ and La_5Ge_3 are almost unaffected by water. The thermal stability of lanthanum germanides in air and ammonia depends directly on their composition. Thus, $LaGe_2$ is more stable with respect to atmospheric oxygen and ammonia than the germanides with a lower germanium content ($LaGe$ and La_5Ge_3).

A systematic investigation of the chemical stability of the rare-earth germanides* showed that they are very unstable in acidic media. Nitric acid, even in a dilution of 1:10, completely decomposes these germanides at $100 \pm 5°C$. The presence of an oxidizing agent (hydrogen peroxide) activates decomposition. Water, 30% hydrogen peroxide solution, and alkali solutions (10, 20, and 40% NaOH) have virtually no effect on rare-earth germanides; these compounds are also resistant to the action of organic solvents.

Actinide Germanides. The compounds $ThGe_3$, α-$ThGe_2$, and β-$ThGe_2$ react vigorously with 5% hydrochloric acid, concentrated hydrofluoric acid, aqua regia, and 10% sodium hydroxide solution at room temperature. The reaction of these phases with 30% hydrogen peroxide, 3 M and 18 M sulfuric acid, and concentrated and 6 M nitric acid is rather slow. No reaction takes place with 85% H_3PO_4 or 0.1% $KMnO_4$ [48].

NONMETAL GERMANIDES

Germanium Nitrides. The compound Ge_3N_4 is stable in air, in water at 100°C, and in boiling sodium hydroxide solution; it is slightly attacked by strong mineral acids.

The compound Ge_3N_4 readily decomposes in water [300].

Germanium Sulfides. Germanium disulfide (GeS_2) is poorly soluble in water, stable with respect to concentrated mineral acids, and highly soluble in hot alkali hydroxides and aluminum hydroxide.

Crystals of germanium sulfide (GeS) are poorly soluble and react slowly with acids and alkalis, even at the boiling point. Powdered germanium sulfide is highly soluble in dilute alkali hydroxides, from which it can be reprecipitated in the form of an amorphous reddish-brown mass by treatment with acids.

Germanium sulfide also dissolves in moderately concentrated hydrochloric acid, liberating hydrogen sulfide. It is poorly soluble in ammonia or ammonium sulfide, but dissolves rather readily in ammonium polysulfide. A white precipitate of germanium disulfide is obtained when these solutions are treated with acids.

Amorphous germanium sulfide reacts well with alkali and sulfide solutions and dilute hydrochloric acid and poorly with phosphoric, sulfuric, and organic acids. It is readily oxidized by hot dilute nitric acid and aqueous solutions of hydrogen peroxide, chlorine, bromine, and potassium permanganate. It is oxidized in air slowly at 350°C and rapidly at higher temperatures, forming GeO_2 and SO_2. The amorphous modification undergoes a transformation to the crystalline modification when held in a nitrogen atmosphere at 460°C for several hours.

---

*K. A. Lynchak, Author's abstract of dissertation [in Russian], Kiev (1968).

Both modifications of GeS react with gaseous chlorine and bromine at room temperature, producing chlorides or bromides of germanium and sulfur and evolving considerable heat. Both modifications are reduced to germanium by hydrogen at high temperatures [3].

The literature contains contradictory data on the oxidation of germanium disulfide [414-417]. The method employed to study oxidation in these investigations was based on determination of the amount of sulfur liberated in the form of SO_2 [414]; in addition, the mass of the sulfide specimen was measured in some studies [415].

Bibikova and Vasilevskaya [414] believe that oxidation of germanium disulfide, during which sulfur dioxide is liberated, begins at 280°C, the rate of the reaction between the GeS_2 and atmospheric oxygen being extremely low. The reaction rate increases at 380°C, but both the rate and extent of oxidation decrease over the temperature range 430-600°C and begin to rise again only at 635°C. On the basis of an x-ray diffraction analysis of the oxidation products, these authors concluded that GeS_2 is oxidized to GeO_2.

Davydov and Diev [416] also feel that germanium dioxide is formed during the oxidation of germanium disulfide by atmospheric oxygen.

Gurovich and Sokolova [327] studied the oxidation of germanium disulfide by the thermogravimetric method. They state that oxidation of GeS_2 begins at 260-280°C with formation of germanium dioxide and that both this compound and germanium sulfide are formed over the temperature range 280-670°C; sulfate formation is most vigorous at 260-380 and 503-575°C.

In the opinion of these authors, the sulfide of tetravalent germanium decomposes at 540-700°C. The oxidation mechanism they propose is not adequately grounded in fact, however, since all their conclusions are based on indirect data (the amount of sulfuric anhydride evolved and the change in specimen mass); the reaction products were not checked for germanium sulfate by x-ray diffraction analysis or chemical phase analysis.

Gurovich and Sokolova [327] studied the oxidation of germanium disulfide by measuring the electrical resistance of specimens in air as a function of temperature and by chemical phase analysis of the residue remaining after oxidation. The resistance curves for germanium disulfide specimens showed that oxidation begins at 250°C. Below 300°C, oxidation of GeS_2 in air results in formation of germanium dioxide and sulfur dioxide and proceeds by the reaction

$$GeS_2 + 3O_2 \rightarrow GeO_2 + 2SO_2.$$

Over the temperature range 300-700°C, oxidation of GeS_2 produces germanium dioxide, sulfur dioxide, and elemental sulfur. Phase analysis established that the S_{SO_2}/S_{O_2} ratio over this temperature range is 0.6, from which it follows that the oxidation reaction conforms to the equation

$$4GeS_2 + 7O_2 \rightarrow 4GeO_2 + 3SO_2 + 5S.$$

In addition to evolution of sulfur dioxide and formation of germanium dioxide, small amounts of complex germanium—sulfur compounds appear during oxidation of germanium disulfide at 700°C and above.

Germanium Selenides. The germanium selenides GeSe and $GeSe_2$ are poorly soluble in hydrochloric and sulfuric acids. Both compounds readily dissolve in hydrochloric acid in the presence of bromine. Hydrogen peroxide oxidizes them to liberate selenium, especially in alkaline media. Nitric acid oxidizes $GeSe_2$ slowly and GeSe considerably more rapidly, producing germanium dioxide and selenious acid. Freshly precipitated GeSe dissolves better in alkali solutions than GeSe obtained by fusion. Substantially more $GeSe_2$ than GeSe dissolves in such solutions [332].

It must be noted that there are almost no data in the literature on techniques for chemical analysis of germanides, since very little research has been done in this area.

CHAPTER 8

APPLICATIONS OF GERMANIUM AND ITS ALLOYS

Wide use of germanium and its alloys began only 5-8 years ago, when techniques were developed for producing these substances in very pure form. Germanium is now one of the main chemical elements on which modern semiconductor technology is based. The remarkable electrophysical properties of germanium have resulted in a rapid increase in its extraction and production, which has permitted its use in the most diverse areas of science and technology. The radio-electronics industry is the principal consumer of germanium. Crystalline semiconductor devices can replace vacuum tubes and provide high reliability, economy, and compactness for complex electronic equipment. Germanium is also utilized in areas of industry outside semiconductor technology [418, 419]. Germanium is widely used in infrared technology. The element itself and glasses based on germanium dioxide transmit infrared radiation. The fact that germanium has a high refractive index in comparison with NaCl (n = 3.458 at a wavelength of 2.0 μ) means that optical surfaces can have a lesser curvature. The fortunate ratio of the refractive indices of germanium and silicon, which is similar to that for flint and crown glasses, makes it possible to combine germanium and silicon lenses in complex objectives. The corrosion resistance of germanium makes infrared optics nonsusceptible to atmospheric and thermal factors.

Germanium is very promising for the metallurgical industry, in the production of various alloys, although the high cost of the element still limits opportunities for its application in this area.

Alloys of germanium with various metals are distinguished by interesting properties and have extremely diverse applications. Many of them have a high resistance to aggressive acidic media, so that they can be used for chemical equipment and protective platings. Thus, copper-germanium alloys (germanium bronzes) are resistant to hydrochloric and sulfuric acids. Bronzes with a content of more than 25% Ge are also resistant to nitric acid and dissolve only in aqua regia.

Another group of acid-resistant alloys that withstand the action of hot and cold mineral acids and acid mixtures are Pd-Ag-Au-Ge alloys containing 10-65% Pd, 15-50% Ag, 20-65% Au, and 0.5-5% Ge. Such alloys, which have high plasticity and are readily machined, are used in the manufacture of various equipment for the chemical industry.

Tin-germanium and antimony-germanium alloys with contents of up to 50 and 11.5% Ge respectively can be used as protective platings [421]. High-quality Sn-Ge platings up to 8 μm thick are produced by electrodeposition from an electrolyte containing 90 g/liter NaOH, 0.45-4.5 g/liter Ge in the form of GeO_2, and 45 g/liter Sn in the form of $SnCl_{14}$; Sb-Ge platings are deposited from an electrolyte containing 180 g/liter NaOH, 100 g/liter Na_2S, 2-10 g/liter Ge, and 10 g/liter Sb. These platings are almost nonporous and have exceptionally good anticorrosion properties. They are stable in sea water, in atmospheres with a high humidity and variable temperature (under "tropical conditions"), and in other corrosive media.

Alloys of gold with 12-13% Ge are used for low-temperature solder in electronics. They are also employed for plating gold and gilt articles. These alloys have comparatively high hardness (200 on the Vickers scale), wet the surface of gold well, and form a dense plating that protects the article against wear. The plating process is facilitated by the comparatively low melting point of the alloys (356°C). Eutectic gold-containing alloys can expand during hardening, so that they can be used for precision castings.

Kochegarov et al. [422] report the results of work on solution formulas and technological regimes for electrochemical deposition of indium to produce indium platings on germanium.

The volt-ampere characteristics obtained for the germanium-indium boundary layer indicate formation of an electron—hole junction; the relatively high d.c. resistance of the germanium slab makes it impossible to make a precise appraisal of the volt-ampere characteristics of the p-n junction. The authors point out that the rectifying properties of the electron—hole junction produced by this method are apparently not as good as those of an electron—hole junction obtained by fusion; however, the fact that indium can be applied to a germanium surface of given shape can be utilized in the production of solid-state circuitry for semiconductor devices having a large indium-electrode surface.

A germanium—indium alloy containing 0.001 at-% In can be used in the manufacture of low-temperature resistance thermometers that operate at the temperature of liquid helium (2-4°K) and have high sensitivity (up to 0.001°K) [420]. Such resistance thermometers also have a low sensitivity to external electric fields and high reading reproducibility.

An alloy containing 74% Al, 21% Ge, 3% Si, and 2% Fe is used as a cathode material for vacuum tubes [423]. It has good secondary emession.

Abeles et al. [424] investigated - the high-temperature thermal conductivity of germanium—silicon alloys of the n and p types. According to their data, Ge—Si alloys can be used as materials for high-temperature thermoelectric power generators by virtue of their low thermal conductivity, high thermal stability and small mass.

Alloys in the Al—Si—Ge system containing up to 58% Ge, and up to 13% Si, and Al are used as solders for aluminum, magnesium, and their alloys. Aluminum can also be soldered with an Sn—Cd—Ge alloy containing 70-80% Sn, 20-30% Cd, and up to 5% Ge.

A silver-germanium alloy has come into use in the fabrication of disk terminals for high-frequency electrical contacts.

Addition of small amounts of germanium substantially improves the properties of an alloy containing about 8% Al and 92% Cu, which can be used as a substitute for gold and other expensive metals in dental prostheses.

Intermetallic germanium compounds are coming into ever wider use. Thus, the compound Mg_2Ge produces an improvement in mechanical properties when added to aluminum alloys [425]. Alloys in the $Al—Mg_2Ge$ system are hardened by quenching and subsequent natural and artificial aging. Naturally and artificially aged alloys containing 2.7% Mg_2Ge conjoin maximum ultimate strength (σ_b = 230-250 and 300-320 MN/m^2 [23-25 and 30-32 kg/mm^2]) with somewhat reduced relative elongation (δ = 22-25 and 10-18%) and transverse necking (ψ = 46-52 and 20-42%).

According to Kroemer et al. [36], the compound Mg_2Ge can be used in the production of devices of the metal-oxide semiconductor type.

The intermetallic compound Cu_3Ge, which has good chemical stability and an attractive color (silvery-white with a bluish tinge), has come into use in the arts and as a protective plating.

On the basis of the results obtained by Wernick et al. [370] in investigating the magnetic properties of MnAlGe, it can be assumed that crystals of this compound, which exhibit strong anisotropy, can be used as sensors to determine the direction of magnetic fields and in various process-automation systems.

According to Brixner [360], tantalum and niobium digermanides and ternary alloys of germanium with these metals can be employed as materials for thermoelements. They have an advantage over sulfides, selenides, and tellurides, since they have a higher melting point, are resistant to oxidation, and can operate at temperatures above 1000°C.

The literature contains reports of the successful use of semiconductive alloys of germanium with titanium, vanadium, chromium, nickel, magnesium, copper, zinc, and other metals.* Addition of these metals gives germanium electronic conductivity, the alloys being used in crystal-detector circuits with high inverse voltage, very low forward resistance, low capacitance, and a twelvefold increase in frequency limit, the latter resulting from the greater increase in current-carrier mobility.

According to Sikirika and Ban [330], ternary alloys in the thorium—transition metal—germanium system are promising for nuclear engineering.

The most important germanium compounds are germanium dioxide (GeO_2), germanium halides, and germanium hydride (GeH_4). They all are of great technological importance, since they are used in the production of ultrapure germanium for semiconductors. Germanium dioxide, being one of the best known vitrifying agents, is extensively used in the manufacture of special glasses with high refractive indices and infrared transparency, enamels, and ceramics. This compound exhibits catalytic properties in some cases and is an activator when added to oxide and halide luminophores. Addition of 0.01-0.02% germanium dioxide to lead and lead alloys reduces their toxicity substantially.

Germanium hydride (GeH_4) is used for producing especially pure metallic germanium.

Further research on the physicochemical and technological properties of germanium and its compounds with metals and nonmetals will obviously make it possible to expand the use of these materials in various areas of modern technology.

---

*Canadian patent no. 518088 [1955]; U. S. patent no. 274506 (1956).

LITERATURE CITED

1. R. J. Jaccodine, J. Electrochem. Soc., 110 (6): 524 (1963).
2. B. A. Krasyuk and A. I. Gribov, Semiconductors: Germanium and Silicon [in Russian], Metallurgizdat (1961).
3. O. Johnson, "Germanium and its compounds," Usp. Khim., 25:1 (1956).
4. A. E. Fersman, Geochemistry [in Russian], Vol. 1, ONTI (1934).
5. E. Gastinger, Fortschritte der chemischen Forschung, Vol. 3, No. 3 (1955).
6. M. S. Sominskii, Semiconductors and Their Application in Engineering [in Russian], Lenizdat (1958).
7. Yu. M. Shashkov, Semiconductor Metallurgy [in Rssian], Metallurgizdat (1960).
8. Yu. V. Shmartsev et al., Refractory Semiconductor Materials, Consultants Bureau, New York (1966).
9. L. Valdes, Proc. IRE, No. 2 (1954).
10. R. Keyes, Phys. Rev., 84:367 (1951).
11. V. S. Vavilov, Usp. Fiz. Nauk, 75 (4):263 (1961).
12. B. Roberts, Superconductive Materials and Some of Their Properties, Report No. 63-RL-3252 M, March, Schenectady, N. Y. (1963).
13. D. Turnbull, J. Appl. Phys., 21:1022 (1950).
14. Van Arkel, Reine Metalle, Berlin (1939).
15. Handbook of Chemistry [in Russian], Goskhimizdat, Moscow (1962).
16. E. Conwell, Proc. IRE, 40 (11):1327 (1952).
17. O. Kubaschewski and E. L. Evans, Metallurgical Thermochemistry, Pergamon Press, New York (1958).
18. E. Greiner, J. Metals, 4 (10):1044 (1952).
19. I. P. Lomashov and B. I. Losev, Germanium in Fossil Coal [in Russian], Izd. An SSSR (1962).
20. M. Straumanis and E. Aka, J. Appl. Phys., 23:330 (1952).
21. A. Crieco and H. Montgomery, Phys. Rev., 86:4 (1952).
22. A. Searcy, J. Am. Chem. Soc., 74 (19):4789 (1952).
23. L. V. Shchipanova and P. V. Gel'd, Izv. Vuzov, Tsvet. Metallurg., No 6, p. 111 (1962).
24. K. Kelley, Bull. Bur. Mines, No. 584 (1960).
25. V. S. Fomenko, Emission Properties of Elements and Chemical Compounds [in Russian], Izd. AN USSR, Kiev (1961).
26. M. M. Kakhana and É. E. Vainshtein, Izv. Akad. Nauk SSSR, Ser. Fiz., Vol 21, No. 10 (1957).
27. H. Glaser, Phys. Rev., 82:616 (1951).
28. K. I. Narbutt, Izv. Akad. Nauk SSSR, Ser. Fiz., 20:780 (1956).
29. É. E. Vainshtein et al., Zh. Eksperim. i Teor. Fiz., 23:593 (1952).
30. É. E. Vainshtein et al., Zh. Eksperim. i Teor. Fiz., 27:521 (1954).
31. Germanium (Collection of Articles) [Russian translation], IL (1955).

32. R. Carreker, J. Met., 8 (2):111 (1956).

33. D. T. Hurd, Introduction to the Chemistry of the Hydrides, Wiley, New York (1952).

34. S. A. Semenkovich, Zh. Fiz. Khim., 39 (9):2232 (1965).

35. N. A. Galaktionova, Hydrogen in Metals [in Russian], Metallurgizdat (1959), p. 116.

36. H. Kroemer et al., J. Appl. Phys., 36 (8):2461 (1965).

37. R. Jaffee et al., Trans. Electrochem. Soc., 89:277 (1946).

38. O. A. Songina, Rare Metals [in Russian], Metallurgizdat (1955).

39. G. A. Kataev and L. N. Rozanova, Problems of Chemical Kinetics and Reactivity: Transactions of the V. V. Kuibyshev Tomsk State University, Chemistry Series [in Russian], Vol. 154 (1962).

40. A. P. Vinogradov, Geochemistry of Rare and Dispersed Elements in Soil [in Russian], 2nd edition, Izd. AN SSSR (1957).

41. H. Onishi, Bull. Chem. Soc. Japan, 29:6 (1956).

42. J. Burton and W. Slichter, Transistor Technology, Vol. 1, Ch. 5, Van Nostrand, Princeton, N. J. (1958).

43. F. Herman, Phys. Rev., 88:1210 (1952).

44. F. Herman and J. Callaway, Phys. Rev., 89:518 (1953).

45. L. Pauling, The Nature of the Chemical Bond, 2nd Edition, Cornell University Press, Ithaca, N. Y. (1945).

46. G. Hagg, Z. Phys. Chem., 12:33 (1931).

47. T. Millner, Z. anorg. allg. Chemie, 292:25 (1957).

48. H. Novotny, Usp. Khim., 27 (8):996 (1958).

49. E. Hohmann, Z. anorg. allg. Chemie, 292:25 (1957).

50. G. V. Samsonov, Silicides and Their Technical Applications [in Russian], Izd. AN Ukr. SSR (1959).

51. G. B. Bokii, Introduction to Crystal Chemistry [in Russian], Izd. MGU (1954).

52. P. Esslinger and K. Schubert, Z. Metallkunde, 48 (3):126 (1957).

53. E. Hellner, Z. Kristallogr., 107:99 (1956).

54. K. Schubert, Z. Naturforsch., 8a:30 (1953).

55. P. Eckerlin and E. Woelfel, Z. anorg. allg. Chemie, 280:321 (1955).

56. G. S. Zhdanov and V. P. Glagoleva, Tr. Inst. Kristallogr. Akad. Nauk SSSR, 9:211 (1954).

57. H. Nowotny et al., Monatsh. Chemie, 88:180 (1957).

58. E. Parthe et al., Monatsh. Chemie, 86:859 (1955).

59. E. Pell, Phys. Chem. Solids, 3 (1-2):74 (1957).

60. E. I. Gladyshevskii and P. I. Kripyadevich, Kristallografiya, 5 (4):574 (1960).

61. G. V. Samsonov, in Rare Alkali Metals [in Russian], Izd. SO AN SSSR, Novosibirsk (1967).

62. R. Busmann, Z. anorg. allg. Chemie. 313:90 (1961).

63. B. Busmann, Naturwiss. 47:82 (1960).

64. R. Schaefer and W. Klemm, Z. anorg. allg. Chemie. 312 (3-4):214 (1961).

65. E. Gebhardt, Z. Metallkunde, 34 (11):255 (1942).

66. H. Spengler, Metall, 8 (23-24):936 (1954).

67. C. Thurmond et al., J. Chem. Phys., 25 (4):799 (1956).

68. V. S. Neshpor and S. N. L'vov, Dokl. Akad. Nauk Ukr. SSR, No. 11, p. 1461 (1962).

69. G. V. Samsonov, Ukr. Khim. Zh., 31:1233 (1965).

70. M. I. Korsunskii and Ya. E. Genkin, Izv. Akad. Nauk SSSR, Ser. Fiz., 28:832 (1964).

71. V. G. Andrianov et al., Izv. Akad. Nauk SSSR, Neorgan. Mat., 2 (11):2064 (1966).

72. L. Brixner, J. Inorg. Nucl. Chem., 25:527 (1963).

73. D. Yost et al., Rare-Earth Elements and Their Compounds [Russian translation], IL (1965).

74. G. Herzberg, Atomic Spectra and Atomic Structure, Dover Press, New York (1944).

75. G. V. Samsonov and V. S. Neshpor, Dokl. Akad. Nauk SSSR. 122:1021 (1958).

LITERATURE CITED

76. Yu. B. Paderno et al., Dokl. Akad. Nauk Ukr. SSR, No. 1, p. 56 (1965).
77. R. Juza and A. Rabenau, Z. anorg. allg. Chemie, 285:212 (1956).
78. R. Heikes and R. Ure, Thermoelectricity, New York (1961).
79. J. Moriguchi and Y. Koga, J. Phys. Soc. Japan, 12 (1):100 (1957).
80. R. Ure et al., Metallurgical Society Conferences, Boston, Mass. (1959).
81. H. Moissan, Der elektrische Ofen, M. Krayn, Berlin (1900).
82. H. Wallbaum, Naturwiss. 32:76 (1944).
83. S. Geller, Acta Cryst. 8 (1):115 (1955).
84. V. S. Lyashenko and V. N. Bykov, Atomnaya Energiya., 8 (2):146 (1960).
85. V. N. Bondarev and G. V. Samsonov, Poroshkovaya Met., No. 1, p. 65 (1964).
86. H. Nowotny et al., Monatsh. Chemie, 91 (2):270 (1960).
87. E. Parthe and J. Norton, Acta Cryst. 11 (1):14 (1958).
88. M. Barbier-Andrieux, Ann Chemie, 10:754 (1955).
89. M. Barbier-Andrieux, Bull. Soc. Franc. Electr., 6 (70):670 (1956).
90. L. Dennis and N. Skow, J. Am. Chem. Soc., 52:2369 (1930).
91. G. I. Oleksiv, in: Development of the Natural and Exact Sciences [in Ukrainian], L'vov Univ., L'vov (1964), p. 76.
92. R. Schwartz and G. Elstner, Z. anorg. Chemie, 217:289 (1934).
93. F. Weibke, Metallwirtsch. 15:299 (1936).
94. W. Hume-Rothery et al., J. Inst. Metals, 66:361 (1940).
95. W. Hume-Rothery et al., Phil. Trans. Roy. Soc., A233 (1): 82 (1934).
96. E. Owen and V. Rowlands, J. Inst. Metals. 66:361 (1940).
97. K. Schubert and G. Brandauer, Z. f. Metallkunde, 43 (7): 262 (1952).
98. K. Schubert and G. Brandauer, Naturwiss. 39 (9): 208 (1952).
99. W. Klemm and H. Westlinning, Z. anorg. allg. Chemie, 245 (4):365 (1941).
100. G. Brauer and J. Tisler, Z. anorg. allg. Chemie, 262 (6):319 (1950).
101. P. Eckerlin et al., Z anorg. allg. Chemie, 281:322 (1955).
102. P. Royen and R. Schwarz, Z. anorg. allg. Chemie, 211:412 (1933).
103. O. Helleis et al., Z. Anorg. allg. Chemie, 320 (1-4):86 (1963).
104. A. Iandelli, Atti Acad. Naz. Lincei Rend. Cl., Sci., Fis. Mat. e Nature, 19 (5):307 (1955).
105. P. Pietrokowdky and P. Duwez, J. Metals, 3 (9):772 (1951).
106. P. Pietrokowsky and P. Duwez, Trans. Am. Inst. Min. Met. Eng., 191:772 (1951).
107. N. V. Ageev and V. P. Samsonov, Zh. Neorgan. Khim., 4 (7):1590 (1959).
108. E. Parthe, Acta Crystallogr., 12:559 (1959).
109. G. Hardy and J. Hulm, Phys. Rev., 89 (4):884 (1953).
110. G. Hardy and J. Hulm, Phys. Rev., 93 (5):1004 (1954).
111. H. Holleck et al., Monatsh. Chemie, 94 (2):497 (1963).
112. W. Rossteutscher and K. Schubert, Z. f. Metallkunde, 56 (11):813 (1965).
113. J. Carpenter and A. Searcy, J. Am. Chem. Soc., 78 (10):2079 (1956).
114. J. Carpenter, J. Phys. Chem., 67:2141 (1963).
115. J. Carpenter and A. Searcy, J. Phys. Chem., 67:2144 (1963).
116. V. N. Bondarev, Poroshkovaya Met., No. 1, p. 35 (1966).
117. I. G. Fakidov and N. N. Grazhdankina, Fiz. Met. i Metalloved., 6 (1):67 (1958).
118. A. Searcy and R. Peavler, J. Am. Chem. Soc., 75 (22):5667 (1953).
119. U. Zeicker et al., Z. f. Metallkunde, 40 (12):433 (1949).
120. U. Zwicker, Z. f. Metallkunde, 42 (8):246 (1951).
121. A. K. Shtol'ts and P. V. Gel'd, Fiz. Met. i Metalloved., 12 (3):462 (1961).
122. A. K. Shtol'ts and P. V. Gel'd, Fiz. Met. i Metalloved., 13 (1):159 (1962).
123. A. K. Shtol'ts et al., Fiz. Met. i Metalloved., 16 (1):130 (1963).
124. A. K. Shtol'ts and P. V. Gel'd, Zh. Fiz. Khim., 38 (8):2067 (1964).
125. A. K. Shtol'ts et al., Zh. Neorgan. Khim., 9 (1):140 (1964).

126. S. D. Margolin and I. G. Fakidov, Fiz. Met. i Metalloved., 9:823 (1960).
127. T. Ohoyama, J. Phys. Soc. Japan, 18:589 (1963).
128. P. Lecocq and A. Michel. Compt. Rend., 253:2335 (1961).
129. F. Laves and H. Wallbaum, Z. anorg. Mineral., 4:17 (1941).
130. H. Pfisterer and K. Schubert, Z. f. Metallkunde, 40 (10):378 (1949).
131. S. Bhan and K. Schubert, Z. f. Metallkunde, 51 (6):327 (1960).
132. E. Stolz and K. Schubert, Chemie der Erde, 22:709 (1962).
133. M. Barbier-Andrieux, Compt. Rend., 241 (3):309 (1955).
134. H. Pfisterer and K. Schubert, Z. f. Metallkunde, 41 (10):358 (1950).
135. K. Ruttewit and B. Masing, Z. Metallkunde, 32:55 (1940).
136. L. Castelliz, Z. Metallkunde, 46:198 (1955).
137. L. Castelliz, Monatsh. Chemie, 84 (4):765 (1953).
138. E. Raub and W. Fritzche, Z. f. Metallkunde, 44:307 (1953).
139. K. Anderko and K. Schubert, Z. f. Metallkunde, 44:307 (1953).
140. G. Weitz et al., Z. f. Metallkunde, 51 (4):239 (1960).
141. E. Parthe, Acta Cryst. 13:868 (1960).
142. N. Baenziger and J. Hegenbarth, Acta Cryst. 17 (5):520 (1964).
143. J. Arbukle and E. Parthe, Acta Cryst. 15:1205 (1962).
144. F. X. Siegel et al., Trans. Met. Soc. AIME. 227:525 (1963).
145. E. I. Gladyshevskii, Theory and Applications of Rare-Earth Metals [in Russian], Izd. Nauka (1964), p. 141.
156. E. I. Gladyshevskii, Zh. Strukt. Khim., 5 (4):568 (1964).
147. M. D. Lyutaya and A. B. Goncharuk, Izv. Akad. Nauk SSSR., Neorgan. Mat., 1 (3):326 (1965).
148. A. Tharp et al., J. Electrochem. Soc., 105:473 (1958).
149. A. Brown, Acta Cryst. 15:652 (1962).
150. A. Brown and J. Norreys, J. Less-Common Metals, 5:302 (1963).
151. Q. Johnson et al., Acta Cryst. 18:131 (1965).
152. E. I. Gladyshevskii et al., Kristallografiya, 9 (8):338 (1964).
153. C. Birchenall, Metal Reviews, 3:235 (1958).
154. V. M. Glazov et al., in: Chemical Bonding in Semiconductors and Solids [in Russian], Izd. Nauka i Tekhnika, Minsk (1965), p. 135.
155. F. Morin and H. Reiss, Phys. Chem. Solids, 3 (3-4): 196 (1957).
156. H. Reiss et al., Bell. Syst. Tech. J., 35:535 (1956).
157. J. Witte and H. Schnering, A. anorg. allg. Chemie, Vol. 327, No. 3-4 (1964).
158. E. Owen and E. Roberts, Phil. Mag., 27:294 (1939).
159. W. Hume-Rothery et al., J. Inst. Metals, 66:191 (1940).
160. H. Nowotny and K. Bachmayer, Monatsh. Chemie, 81:669 (1950).
161. C. Fuller and J. Struthers, Phys. Rev., 87:526 (1952).
162. C. Fuller et al., Phys. Rev., 93:1182 (1954).
163. C. Thurmond, J. Phys. Chem., 57 (8):827 (1953).
164. C. Thurmond and J. Struthers, J. Phys. Chem., 57:831 (1953).
165. R. Hodkinson, Phil. Mag., 46:410 (1955).
166. J. Reynolds and W. Hume-Rothery, J. Inst. Metals, 85 (4):119 (1956/57).
167. A. Smith, J. Inst. Metals, 80 (9):477 (1951/52).
168. T. Briggs et al., J. Phys. Chem., 33:1080 (1929).
169. A. A. Bugai et al., Zh. Tekhn. Fiz., 27 (8):1671 (1957); 27 (1):210 (1957).
170. A. E. Vol, Structure and Properties of Binary Metallic Systems [in Russian], Vol. 2, Fizmatgiz (1962).
171. Y. Weiling, J. Phys. Chem. Solids, 18 (2-3):162 (1961).
172. R. Jaffee et al., Metals Technology, Vol. 12, No. 4 (1945).
173. R. Jaffee et al., Trans. Am. Inst. Min. Met. Eng., 161:366 (1945).

174. V. E. Kosenko, Fiz. Tverd. Tela, 1 (10):1622 (1959).
175. E. Owen and E. J. O'Donnel-Roberts, J. Inst. Metals, 71:213 (1945).
176. R. V. Skolozdra et al., Zh. Strukt. Khim., 6 (3):473 (1965).
177. A. Kaufmann et al., Trans. Amer. Soc. Metals, 42:785 (1950).
178. Yu. I. Belyaev and V. A. Zhidkov, Fiz. Tverd. Tela, 3 (1):182 (1961).
179. G. Raynor, J. Inst. Metals, 66:403 (1940).
180. E. M. Savitskii and V. V. Baron, Izv. Sekt. Fiz.-Khim. Analiza, 27:86 (1956).
181. E. M. Savitskii et al., Papers Presented at a Conference on Phase Diagrams [in Russian],
        Izd. AN SSSR, Moscow (1956), p. 37.
182. J. Dismukes et al., J. Appl. Phys., 35 (10):2899 (1964).
183. E. M. Savitskii et al., Zh. Neorgan. Khim., 3 (3):763 (1958).
184. U. Winkler, Helv. Phys. Acta, 28 (7):633 (1955).
185. V. N. Eremenko and G. M. Lukashenko, Izv. Akad. Nauk SSSR. Neorgan. Mat., 1 (8):1296
        (1965).
186. V. N. Eremenko and G. M. Lukashenko, Zh. Neorgan. Khim., 9 (7):1552 (1964).
187. V. N. Eremenko and G. M. Lukashenko, Ukr. Khim. Zh., 28:462 (1962).
188. V. G. Fomin et al., Kristallografiya, 9 (2):227 (1964).
189. G. F. Voronin and A. M. Evseev, Zh. Fiz. Khim., 33 (9):2024 (1959).
190. T. Edwards, Phil. Mag., 2:15 (1926).
191. E. I. Gladyshevskii, Dokl. Akad. Nauk Ukr. SSR, No. 3, p. 294 (1959).
192. E. I. Gladyshevskii et al., Kristallografiya. 6:267 (1961).
193. E. I. Gladyshevskii, Dokl. Akad. Nauk Ukr. SSR, No. 2, p. 209 (1964).
194. E. I. Gladyshevskii et al., in: Titanium and Its Alloys [in Russian], Vol. 10, Izd. AN SSSR
        (1963).
195. F. Spedding and A. Daane, The Rare Earths, New York (1961), p. 237.
196. A. Iandelli, Atti. Acad. Naz. Lincei, 6:727 (1949).
197. A. Iandelli and R. Ferro, Ann. Chemie (Roma), 42:598 (1952).
198. O. Schob and E. Parthe, Monatsh. Chemie, 95 (6):1466 (1964).
199. O. Schob and E. Parthe, Acta Cryst. 19 (2):214 (1965).
200. E. Parthe et al., Naturwiss. 52 (7):155 (1965).
201. E. I. Gladyshevskii and V. V. Burnashova, Izv. Akad. Nauk SSSR, Neorgan. Mat., 1 (9):
        1508 (1965).
202. E. I. Gladyshevskii, Zh. Strukt. Khim., 5 (6):919 (1964).
203. E. I. Gladyshevskii and N. S. Ugrin, Dokl. Akad. Nauk Ukr. SSR, No. 1, p. 1326 (1965).
204. J. Moriarty and N. Baenziger, Acta Cryst. 14:946 (1961).
205. J. Moriarty et al., Acta Cryst. 19 (2):285 (1965).
206. G. Smith et al., Acta Cryst. 18:1085 (1965).
207. B. Matthias et al., Phys. Rev., 112:89 (1958).
208. M. McQuillan, J. Inst. Metals, 83 (11):485 (1954).
209. R. Goldhoff et al., Trans. Am. Soc. Metals, 45:941 (1953).
210. N. V. Ageev et al., Zh. Neorgan. Khim., 4 (8):1864 (1959).
211. N. V. Ageev and V. P. Samsonov, Zh. Neorgan. Khim., 4 (7):1590 (1959).
212. I. I. Kornilov, Zh. Neorgan. Khim., 3 (2):364 (1958).
213. B. Lustman and F. Kerze, The Metallurgy of Zirconium, McGraw-Hill, New York (1955).
214. O. Carlson et al., Trans Am. Soc. Metals, 45:941 (1953).
215. J. Smith and D. Bailey, Acta Cryst. 10 (4):341 (1957).
216. H. Voellenkle et al., Monatsh. Chemie, 95 (6):1544 (1964).
217. O. Holleck et al., Monatsh. Chemie, 96 (2):570 (1965).
218. M. Sarachik et al., Canad. J. Phys., 41 (10):1542 (1963).
219. H. Nowotny et al., J. Phys. Chem., 60 (5):677 (1956).
220. V. L. Zagryazhskii et al., Izv. Akad. Nauk SSSR, Neorgan. Mat., 1 (11):1917 (1965).

221. A. Searcy et al., J. Am. Chem. Soc., 74 (2):566 (1952).
222. A. Searcy et al., J. Am. Chem. Soc., 76 (21):5287 (1954).
223. E. Raub and W. Plate, Z. f. Metallkunde, 42 (3):76 (1951).
224. P. Stecher et al., Monatsh. Chemie, 94 (6):1154 (1963).
225. A. Brown, Nature, 206 (4983):502 (1965).
226. R. Peavler and A. Searcy, J. Am. Chem. Soc., 78 (10):2076 (1964).
227. V. S. Neshpor and S. S. Ordah'yan, Poroshkovaya Met., No. 1, p. 23 (1964).
228. J. Downing and D. Cubicciotti, J. Am. Chem. Soc., 73 (8):4025 (1951).
229. K. Yasukochi et al., J. Phys. Soc. Japan, 15:932 (1960).
230. V. N. Novogrudskii and I. G. Fakidov, Zh. Eksp. i Teor. Fiz., No. 1 (7), p. 40 (1964).
231. I. K. Kikoin et al., Dokl. Akad. Nauk SSSR, 125:1011 (1950).
232. A. Vrzhetsiono, Fiz. Met. i Metalloved., 18 (6):944 (1964).
233. A. Vrzhetsiono, Fiz. Met. i Metalloved., 14 (2):182 (1962).
234. S. D. Margolin and I. G. Fakidov, Fiz. Met. i Metalloved., 5 (2):368 (1957).
235. I. G. Fakidov et al., Izv. Akad. Nauk SSSR, Ser. Fiz. 20 (12):1509 (1956).
236. I. G. Fakidov and Yu. N. Tsiovkin, Fiz. Met. i Metalloved., 7 (5):685 (1959).
237. G. V. Samsonov, Beryllides [in Russian], Naukova Dumka, Kiev (1966).
238. H. Wallbaum, Z. f. Metallkunde, 35:218 (1943).
239. F. Wever, Naturwiss. 17:304 (1929).
240. M. I. Chiragov and A. G. Talybov, Kristallografiya, 10 (3):409 (1965).
241. L. D. Dudkin, Fiz. Tverd. Tela, 2 (3):397 (1960).
242. L. D. Dudkin and V. I. Vaidanich, Fiz. Tverd. Tela, 2 (3):404 (1960).
243. J. Forsyth et al., Phil. Mag., 10 (106):713 (1964).
244. E. Kren and P. Szabo, Phys. Letters, 11 (3):215 (1964).
245. A. K. Shtol'ts et al., Izv. Vuzov, Tsvetn. Metal., No. 2, p. 120 (1965).
246. L. A. Filatova et al., Mechanism of the Reaction of Metals with Gases [in Russian], Izd. Nauka (1964), p. 160.
247. O. Kubaschewski and B. E. Hopkins, Oxidation of Metals and Alloys [Russian translation], IL (1955), p. 138. [English edition (2nd Ed.): Butterworth, London (1962).]
248. W. Koster and E. Horn, Z. f. Metallkunde, 43:333 (1952).
249. W. Pearson and L. Thompson, Canad. J. Phys., 35:349 (1957).
250. F. Maesen and J. Brenkman, Phillips Res. Repts., 9 (3):225 (1954).
251. R. Hall, Phys. Chem. Solids, 3 (1-2):63 (1957).
252. N. N. Zhuravlev and G. S. Zhdanov, Kristallografiya, 1:205 (1956).
253. B. Matthias, Phys. Rev., 91 (2):413 (1953).
254. K. Schubert and K. Anderko, Naturwiss. 39:351 (1952).
255. K. Schubert et al., Naturwiss. 50 (2):41 (1963).
256. B. Frost and J. Maskrey, J. Inst. Metals, 82 (4):171 (1953).
257. E. S. Makarov and V. N. Bykov, Kristallografiya, 4 (2):183 (1959).
258. A. A. Bochvar et al., Proceedings of the 2nd International Conference on the Peaceful Uses of Atomic Energy, Geneva, 1958. Papers Presented by Soviet Scientists. Nuclear Fuels and Reactor Metals [in Russian], Atomizdat (1959), p. 376.
259. A. S. Coffinberry and F. H. Ellinger, Proceedings of the 1st International Conference on the Peaceful Uses of Atomic Energy, Geneva, 1955. Reactor Technology and Chemical Processing of Nuclear Fuels [Russian translation], Vol. 9, Khimizdat, Leningrad (1958), p. 174.
260. L. Lagrenaudie, J. Chim. Phys. et Phys. Chim. Biol., 50 (5):352 (1953).
261. G. V. Samsonov et al., Boron, Its Compounds, and Its Alloys [in Russian], Izd. AN SSSR (1960), p. 203.
262. M. Struge, Proc. Phys. Soc., 73 (Pt. 2):470, 320 (1959).
263. H. Stoehr and Z. Klemm, Z. anorg. allg. Chemie, 241:305 (1939).

264. H. Stoehr and W. Klemm, Z. anorg. allg. Chemie, 244:205 (1940).
265. H. Axon and W. Hume-Rothery, Proc. Roy. Soc. (London), A1 93:1 (1948).
266. E. A. Boom, Izv. Akad. Nauk SSSR. Otd. Khim. Nauk, No. 3, p. 317 (1947).
267. E. A. Boom, Dokl. Akad. Nauk SSSR, 84 (4):697 (1952).
268. O. Sherby et al., J. Metals, 3 (8):643 (1951).
269. A. Robinson and J. Dorn, J. Metals, 3 (6):457 (1951).
270. W. Klemm et al., Z. anorg. allg. Chemie, 256:239 (1948).
271. E. Greiner and P. Breidt, J. Metals, 3 (6):457 (1951).
272. E. Greiner and P. Breidt, Trans. AIME, 203:187 (1955).
273. G. Grube and G. Hutt, Naturforschung und Medizin in Deutschland in 1939–1946, Anorg. Chemie, 25:257.
274. N. DeRoche, Z. f. Metallkunde, 48:59 (1957).
275. P. Keck and J. Broder, Phys. Rev., 90 (4):521 (1953).
276. I. A. Sheka et al., Gallium [in Russian], Gostekhizdat USSR, Kiev (1963), p. 213.
277. J. Roschen and C. Thornton, J. Appl. Phys., 29:923 (1958).
278. Yu. P. Khukhryanskii, Fiz. Tverd. Tela. 6 (5):1557 (1964).
279. Ya. A. Ugai and Yu. P. Khukhryanskii, Fiz. Tverd. Tela. Vol. 6, No. 3 (1964).
280. F. Trumbore, Bell Syst. Tech. J., 39:205 (1960).
281. V. M. Kozlovskaya and R. N. Rubinshtein, Fiz. Tverd. Tela, 3:2354 (1961).
282. V. I. Tagirov and A. A. Kuliev, Fiz. Tverd. Tela, 4 (1):274 (1962).
283. R. Scace and G. Slack, J. Chem. Phys., 30:6 (1959).
284. M. Barbier-Andrieux. Comptes Rendus, 242:2352 (1956).
285. F. Hassion et al., J. Phys. Chem., 59:1118 (1955).
286. E. Johnson and S. Christian, Phys. Rev., 95 (2):560 (1934).
287. Yu. B. Kuz'ma et al., Zh. Strukt. Khim., 3 (2):157 (1962).
288. A. Levitas et al., Phys. Rev., 95:846 (1954).
289. C. Wang and B. Alexander, Acta Metallurgica, 3 (5):515 (1955).
290. C. Wang and B. Alexander, AIME Symposium on Semiconductors, February 15–18, New York (1954).
291. W. Gueriler and M. Pirani, Metallwirtsch., 16:749 (1937).
292. V. M. Vasilevskaya and E. G. Miselyuk, Ukr. Fiz. Zh., 3 (1):71 (1958).
293. L. Dowell and K. Lark-Horovitz, Proc. Indiana Acad. Sci., 56:237 (1946).
294. W. Johnson, Proc. Indiana Acad. Sci., 57:180 (1947).
295. F. Trumbore, J. Electrochem. Soc., 103:597 (1956).
296. R. Rogers and J. Fydell, J. Electrochem. Soc., 100 (4):161 (1953).
297. N. L. Pokrovskii and L. S. Tissen, Zh. Fiz. Khim., 34 (6):1238 (1960).
298. T. Briggs and W. Benedict, J. Phys. Chem., 34:173 (1930).
299. O. Shvarts, Usp. Khim., 5 (10):1448 (1936).
300. A. Busev, Uch. Zap. Leningradsk. Gos. Ped. Inst., No. 23, p. 303 (1940).
301. O. Johnson, J. Am. Chem. Soc, 52:5160 (1930).
302. O. Johnson and R. Ridgely, J. Am. Chem. Soc., 56:2395 (1934).
303. R. Juza and H. Hahn, Z. anorg. allg. Chemie, 27:32 (1939).
304. R. Juza and H. Hahn, Z. anorg. allg. Chemie, 244:125 (1940).
305. W. Leslie et al., J. Metals, 4 (2):204 (1952).
306. O. Johnson, Chemical Reviews, 51 (3):431 (1952).
307. J. Bryden, Acta Cryst. 15:167 (1962).
308. K. Schubert et al., Naturwiss., 41 (19:488 (1954).
309. V. M. Glazov and S. N. Chizhevskaya, Zavod. Lab., 26 (6):720 (1960).
310. N. V. Kolomoets et al., Fiz. Tverd. Tela, 6:706 (1964).
311. W. Bues and H. Wartenberg, Z. anorg. allg. Chem., 266 (6):281 (1951).
312. N. P. Alimarin and B. N. Ivanov-Émin, Zh. Prikl. Khim., 17 (4–5):204 (1944).

313. E. Gastinger, Naturwiss. 42 (4):95 (1955).

314. E. Gastinger, Z. anorg. allg. Chemie, 285:103 (1956).

315. M. Hoch and H. Johnston, J. Chem. Phys., 22 (8):1376 (1954).

316. E. Candidus and D. Tuomi, J. Chem. Phys., 23 (3):588 (1955).

317. V. I. Davydov, Zh. Neorgan. Khim., 2 (7):1460 (1957).

318. A. Pflugmacher and I. Kellermann, Angew. Chemie, 68:374 (1956).

319. O. Rosner, Z. f. Metallkunde, 48 (3):137 (1957).

320. F. Trumbore et al., J. Chem. Phys., 24 (5):112 (1956).

321. W. Zachariasen, Phys. Rev., 49:884 (1936).

322. W. Zachariasen, J. Chem. Phys. 4:618 (1936).

323. N. P. Diev and V. I. Davydov, Izv. Vostochn. Fil. Akad. Nauk SSSR. No. 7, p. 60 (1957).

324. S. V. Ivanova and V. I. Davydov, Zh. Neorgan. Khim., 3 (4):1060 (1958).

325. H. Kenworthy et al., J. Metals, 8 (5):682 (1956).

326. R. Barrow et al., Trans. Faraday Soc., 51 (11):1480 (1955).

327. N. A. Gurovich and É. I. Sokolova, Zh. Neorgan. Khim., 9 (7):1537 (1964).

328. V. I. Davydov and N. P. Diev, Zh. Neorgan. Khim., 2 (9):2003 (1957).

329. E. Shimazaki and T. Wada, Bull. Chem. Soc. Japan, 29 (3):294 (1956).

330. M. Sikirika and Z. Ban, Croat, Chem. Acta, 36 (3):151 (1964).

331. T. Yabumoto, J. Phys. Soc. Japan, 13:559 (1959).

332. B. N. Ivanov-Émin, Zh. Organ. Khim., 10 (21):1813 (1940).

333. Liu Ch'iun-Hua et al., Dokl. Akad. Nauk SSSR, 146 (5):1092 (1962).

334. Liu Ch'iun-Hua et al., Zh. Neorgan. Khim., 7 (5):963 (1962).

335. W. Klemm and G. Frischmuth, Z. anorg. allg. Chemie, 218:249 (1934).

336. K. Schubert and H. Fricke, Z. f. Metallkunde, 44 (10):457 (1953).

337. L. E. Shelimova et al., Zh. Neorgan. Khim., 10 (5):1200 (1965).

338. J. McHugh and W. Teller, Trans AIME, 218 (1):186 (1960).

339. K. Schubert and H. Fricke, Naturwiss. 38:781 (1951).

340. N. V. Kolomoets et al., Fiz. Tverd. Tela, 5:10 (1963).

341. A. A. Andreev et al., Fiz. Tverd. Tela. 7 (2):652 (1965).

342. M. Knudsen, Ann. Phys., 28:75 (1909).

343. H. Stadelmaier and A. Fraker, Z. f. Metallkunde, 53 (1):48 (1962).

344. T. A. Badaeva and R. I. Kuznetsova, Tr. Inst. Metallurgii Akad. Nauk SSSR, No. 3, p. 216 (1958).

345. V. M. Glazov et al., Izv. Akad. Nauk SSSR. OTN, Metallurgiya i Toplivo, No. 4, p. 153 (1959).

346. V. M. Glazov and Liu Ch'iun-Hua, Izv. Akad. Nauk SSSR, OTN, Metallurgiya i Toplivo, No. 4, p. 150 (1960).

347. V. M. Glazov et al., Zh. Neorgan. Khim., 7 (3):576 (1962).

348. V. M. Glazov and Liu Ch'iun-Hua, Zh. Neorgan. Khim., 7 (3):582 (1962).

349. V. M. Glazov et al., Tr. Inst. Metallurgii A. A. Baikova, No. 14 (1963).

350. V. S. Zemskov et al., Izv. Akad. Nauk SSSR, Metallurgiya i Toplivo, OTN, No. 6, p. 149 (1961).

351. B. G. Zhurkin et al., Izv. Akad. Nauk SSSR, OTN, Metallurgiya i Toplivo, No. 4, p. 156 (1959).

352. N. Kh. Abrikosov et al., Zh. Neorgan. Khim., 7 (4):831 (1962).

353. E. M. Savitskii et al., Zh. Neorgan. Khim., 9 (8):2045 (1964).

354. V. M. Glazov and G. L. Malyutina, Zh. Neorgan. Khim., 8 (10):2372 (1963).

355. G. Zwingmann, Z. f. Metallkunde, 55 (4):192 (1964).

356. R. Rayson et al., Acta Metallurgica, 59:125 (1959).

357. V. M. Glazov and D. A. Petrov, Izv. Akad. Nauk SSSR, OTN, No. 4, p. 125 (1958).

358. G. Raynor amd T. Massalski, J. Inst. Metals, 84 (3):66 (1955).

359. H. Holleck et al., Monatsh. Chemie, 93 (5):996 (1962).

360. L. Brixner, J. Inorg. Nucl. Chem., 25:783 (1963).

361. N. V. Ageev and V. F. Shamrai, Izv. Akad. Nauk SSSR, Neorgan. Mat., 1 (8):1277 (1965).

362. N. Kh. Abrikosov et al., Dokl. Akad. Nauk SSSR, 123 (2):279 (1958).

363. H. Stadelmaier and W. Hardy, Z. f. Metallkunde, 52 (6):391 (1961).

364. G. Zwingmann, Metall, 18 (7):726 (1964).

365. P. Stecher et al., Monatsh. Chemie, 94 (3):549 (1963).

366. G. Giesecke and H. Pfister, Acta Cryst. 14:1289 (1961).

367. A. Raman and K. Schubert, Z. Metallkunde, 56 (1):44 (1965).

368. K. A. Gschneider, Rare Earth Alloys, New York (1961).

369. W. Freundlich and M. Chartier, Bull. Soc. Chim. France, 1:63 (1961).

370. J. Wernick et al., J. Appl. Phys., 32 (11):2495 (1961).

371. G. B. Bokii et al., Zh. Strukt. Khim., 2 (1):75 (1961).

372. W. Jeitschko et al., Monatsh. Chemie, Vol. 94, No. 5 (1963).

373. W. Jeitschko et al., Monatsh. Chemie, Vol. 94, No. 6 (1963).

374. W. Jeitschko et al., Monatsh. Chemie, Vol 95, No. 1 (1964).

375. W. Jeitschko et al., Monatsh. Chemie, 95 (4–5):1242 (1964).

376. W. Rieger et al., Monatsh. Chemie, 96 (1):98 (1965).

377. J. A. Perri, Dissertation, Brooklyn Polytechnic Institute (1958).

378. N. Kh. Abrikosov and G. T. Danilova-Dobryakova, Izv. Akad. Nauk SSSR, Neorgan. Mat., 1 (2):204 (1965).

379. N. Kh. Abrikosov and G. T. Danilova-Dobryakova, Izv. Akad. Nauk SSSR, Neorgan. Mat., 1 (1):57 (1965).

380. H. Nowotny and F. C. Benesovsky, Monatsh. Chemie, 92:365 (1961).

381. M. Yu. Teslyuk and E. E..Cherkashin, Dokl. Akad. Nauk Ukr. SSR, 9:1172 (1961).

382. M. Yu. Teslyuk and V. Ya. Markiv, Kristallografiya, 7 (1):128 (1962).

383. M. Yu. Teslyuk et al., Zh. Strukt. Khim., 5 (3):392 (1964).

384. O. Folbert and H. Pfister, Acta Cryst. 14:325 (1961).

385. C. Goodman, Nature, 179:828 (1957).

386. H. Pfisterer, Acta Cryst. 11:221 (1958).

387. N. A. Goryunova et al., Fiz. Tverd. Tela, 5 (7):2031 (1963).

388. N. A. Goryunova et al., Izv. Akad. Nauk SSSR, Neorgan. Mat., 1 (6):885 (1965).

389. E. Fruchart and R. Fruchart, C. R. Acad. Sci. Paris, 258 (11):3032 (1964).

390. E. E. Cherashin et al., Zh. Neorgan. Khim., 3 (3):650 (1958).

391. H. U. Schuster, Naturwiss. 52 (23):639 (1965).

392. E. I. Gladyshevskii et al., Kristallografiya. 6 (5):769 (1961).

393. B. K. Vul'f, Usp. Khim., 29 (6):775 (1960).

394. M. Yu. Teslyuk and G. I. Oleksiv, Dokl. Acad. Nauk Ukr. SSR, No. 10 (1965).

395. R. Juza and W. Schulz, Z. anorg. allg. Chemie, 275:65 (1954).

396. L. Huetter and H. Stadelmeier, Z. Metallurgica, 11 (12:1255 (1963).

397. W. B. Pearson, Handbook of Lattice Spacing and Structures of Metals and Alloys, Pergamon Press, London (1958).

398. H. W. King et al., Acta Metallurgica, 11 (12):1355 (1963).

399. W. Burkhardt and K. Schubert, Z. f. Metallkunde, 50 (4):196 (1959).

400. L. S. Palatnik et al., Kristallografiya, 6 (6):960 (1961).

401. L. S. Palatnik et al., Fiz. Tverd. Tela, 4 (6):1431 (1962).

402. L. I. Berger and A. É. Balanevskaya, Fiz. Tverd. Tela, 6 (5):1311 (1964).

403. L. K. Borusevich, in: Development of the Natural and Exact Sciences [in Ukrainian], L'vov Univ., L'vov (1964), p. 78.

404. B. T. Matthias, Phys. Rev., 92 (4):874 (1953).

405. B. T. Matthias et al., Rev. Mod. Phys., Vol 35 (1963).

406. E. I. Gladyshevskii et al., Dokl. Akad. Nauk Ukr. SSR, No. 4, p. 481 (1962).
407. M. Elli and A. Mugnoli, Atti Acad. Naz. Lincei Rend. Cl. Sci. Fis., Mat. e Nature, 33: (5):315 (1962/63).
408. V. Ya. Markiv et al., Izv. Akad. Nauk SSSR, Neorgan. Mat., 1 (6):890 (1965).
409. H. Hahn and H. Schulze, Naturwiss. 52 (14):426 (1965).
410. H. Stadelmaier and F. Lee, Metall, 18 (2):111 (1964).
411. W. Zachariasen, Phys. Rev., 40:917 (1932).
412. Analysis of High-Melting Compounds [in Russian], Metallurgizdat (1962).
413. Methods for Analysis of Rare Elements [in Russian], Izd. AN SSSR (1961).
414. V. I. Bibikova and I. I. Vasilevskaya, in: Scientific Works from the State Scientific Research and Planning Institute of the Rare Metals Industry [in Russian], Vol. 1, Metallurgizdat (1959).
415. M. D. Galimov and A. I. Okunev, Izv. Vuzov, Ser. Tsvetn. Metal., No. 3, p. 94 (1961).
416. V. I. Davydov and N. P. Diev, Tr. Inst. Metal. Ural'sk. Fil. Akad. Nauk SSSR, No. 4, p. 23 (1958).
417. R. B. Bernstein and D. Cubiciotti, J. Am. Chem. Soc., 73:4112 (1951).
418. N. P. Sazhin, Germanium and Its Applications: Chemistry and Industry [in Russian], Vol.1 (1956), p. 5.
419. N. M. El'khones et al., Germanium and Its Compounds: Present and Prospective Applications [in Russian], Metallurgizdat (1959).
420. E. S. Makarov et al., Izv. Akad. Nauk SSSR, Neorgan. Mat., 2 (12):2116 (1966).
421. S. I. Sklyarenko et al., Izv. Vuzov, Tsvetn. Metal., No. 2, p. 129 (1962).
422. V. M. Kochegarov et al., Pribory i Tekhnika Eksperim., No. 3, p. 187 (1963).
423. C. Fink and V. Dokras, J. Electrochem. Soc., 95:80 (1949).
424. B. Abeles et al., Phys. Rev., 125 (1):44 (1962).
425. I. N. Fridlyander and A. M. Zakharov, Deformable Aluminum Alloys [in Russian], Oborongiz (1961).
426. K. Schubert et al., Naturwiss. 51 (12):287 (1964).
427. P. Popper and S. Ruddlesden, Nature, 179 (4570):1129 (1957).
428. K. Van Con, C. R. Acad. Sci. Paris, 260:111 (1965).
429. L. E. Shelimova et al., Izv. Akad. Nauk SSSR, Neorgan. Mat., 2 (12):2103 (1966).
430. K. Kanematsu, J. Phys. Soc. Japan, 18:800 (1963).
431. R. Ciszewski, Physica Status Solidi, 3:1999 (1963).
432. R. Labotz et al., J. Electr. Soc., 110 (2):127 (1963).
433. E. Gangeberger et al., Monatsh. Chemie, 96:1658 (1965).
434. D. Hohnke and E. Parthe, Acta Cryst. 20 (4):572 (1966).
435. A. G. Tharp et al., Acta Cryst., 20 (4):583 (1966).
436. P. I. Fedorova and V. A. Molochko, Izv. Akad. Nauk SSSR, Neorgan. Mat., 2 (10):1870 (1966).
437. A. É. Balanevskaya et al., Izv. Akad. Nauk SSSR, Neorgan. Mat., 2 (5):810 (1966).
438. V. V. Burnashova and E. I. Gladyshevskii, Izv. Akad. Nauk SSSR, Neorgan. Mat., 2 (5):944 (1966).
439. V. M. Glazov et al., Izv. Akad. Nauk SSSR, Neorgan. Mat., 2 (5):850 (1966).
440. D. Hohnke and E. Parthe, Acta Cryst. 21 (3):435 (1966).
441. G. S. Smith and A. G. Tharp, Nature, 210 (5041):1148 (1966).
442. V. L. Zagryazhskii et al., Poroshkovaya Met., No. 8, p. 55 (1966).
443. E. A. Zhurakovskii, Poroshkovaya Met., No. 8, p. 70 (1966).
444. E. S. Makarov et al., Izv. Akad. Nauk SSSR, Neorgan. Mat., 2 (3):485 (1966).
445. V. L. Zagryazhskii, Izv. Akad. Nauk SSSR, Neorgan. Mat., 2 (2):251 (1966).
446. J. C. Wooley, J. Electrochem. Soc., 112 (9):906 (1965).
447. O. I. Bodak et al., Izv. Akad. Nauk SSSR, Neorgan Mat., 2 (12):2151 (1966).

448. N. E. Alekseevskii et al., Izv. Akad. Nauk SSSR, Neorgan. Mat., 2 (12):2156 (1966).

449. V. M. Svechnikov and F. G. Kobzenki, Dokl. Akad. Nauk Ukr. SSR, Ser. A, No. 3, p. 269 (1967).

450. G. M. Kuznetsov, Izv. Vuzov, Tsvetn. Metal., No. 4, p. 95 (1966).

451. K. Bushow and J. Fast, Physica Status Solidi, 16 (2):467 (1966).

452. Kazuko Senizawa, J. Phys. Soc. Japan, 21 (6):1137 (1966).

453. É. E. Vainshtein et al., Izv. Akad. Nauk SSSR, Neorgan. Mat., 3 (4):644 (1967).

454. G. Smith et al., Acta Cryst. 22 (2):269 (1967).

455. N. N. Serebrennikov et al., Izv. Vuzov, Tsvetn. Metal., No. 3, p. 103 (1966).

456. K. Jain and S. Bhan, Trans. Indian Inst. Metals, 19 (3):49 (1966).

457. V. N. Bondarev and G. V. Samsonov, Poroshkovaya Met., No. 6, p. 52 (1966).

458. V. S. Zemskov et al., Izv. Akad. Nauk SSSR, Neorgan. Mat., 3 (4):705 (1967).

459. M. D. Lyutaya et al., Zh. Neorgan. Khim., 9 (7):1529 (1964).

460. J. Becker and E. Symes, J. Appl. Phys., 36 (3):1000 (1965).

461. J. Mayer, J. Less-Common Metals, 12 (1):46 (1967).

462. W. Jeitschko and E. Parthe, Acta Cryst. 22 (3):417 (1967).

463. H. Haschke et al., Monatsh. Chemie, 97 (5):1452 (1966).

464. V. N. Bondarev et al., Izv. Akad. Nauk SSSR, Neorgan. Mat., 3 (4):707 (1967).

465. E. S. Makarov et al., Izv. Akad. Nauk SSSR, Neorgan. Mat., 3 (2):329 (1967).

466. N. Kh. Abrikosov and S. N. Chizhevskaya, Izv. Akad. Nauk SSSR, Neorgan. Mat., 4 (7):1170 (1968).

467. B. M. Rud' and K. A. Lynchak, Poroshkovaya Met., No. 3, p. 72 (1968).

468. G. V. Samsonov, U. B. Paderneau (Yu. B. Paderno), and B. M. Rud (B. M. Rud'), Revue International des Hautes Temperatures et des Refractairies, No. 12 (1967).

469. G. V. Samsonov, Yu. B. Paderno, and B. M. Rud', Izv. Vuzov, Ser. Fiz., No. 7, p. 105 (1968).

470. G. V. Samsonov, Ukr. Khim. Zh., 33:763 (1967).

471. E. Parthe, Colloque International du CNRS sur les Derives Semimetalliques, Orsay (1965).

472. W. Fife, Introduction to Solid-State Geochemistry [Russian translation], Izd. Mir, Moscow (1967).

473. G. K. Averkieva et al., Izv. Akad. Nauk SSSR, Neorgan. Mat., 4 (7):1064 (1968).

474. V. P. Panus et al., Izv. Akad. Nauk SSSR, Neorgan. Mat., 4 (6):885 (1968).

475. V. P. Panus et al., Izv. Akad. Nauk SSSR, Neorgan. Mat., 6 (6):889 (1968).

476. P. R. Sahm and L. H. Gnau, Z. Metallkunde, 59:137 (1968).

477. M. Richardson, Acta Chem. Scand., 21:753 (1967).

478. R. Merlo and M. L. Formasini, J. Less-Common Metal., No. 2, p. 94 (1967).

479. L. V. Shchipanova et al., Izv. Vuzov, Tsvetn. Metal., No. 2, p. 94 (1967).

480. H. Wolfsgruber et al., Monatsh. Chemie, 98:2403 (1967).

481. H. Voellenkle et al., Z. Kristallogr., 124 (1-2):9 (1967).

482. L. K. Borusevich and E. I. Gladyshevskii, Izv. Akad. Nauk SSSR, Neorgan. Mat., 4 (6):909 (1968).

483. K. A. Lynchak and T. Ya. Kosolapova, Poroshkovaya Met., No. 11, p. 92 (1967).

484. K. A. Lynchak, T. Ya. Kosolapova, and Yu. B. Kuz'ma, Poroshkovaya Met., No. 2, p. 54 (1968).

485. D. P. Shashkov, Izv. Akad. Nauk SSSR, Metally, No. 4, p. 114 (1968).

486. V. L. Zagryazhskii et al., Izv. Vuzov, Fiz., No. 5, p. 46 (1968).

487. H. H. Stadelmaier and G. Hofer, Monatsh. Chemie, 98 (2):408 (1967).

488. Omitted from Russian edition.

489. E. Parthe, Colloq. Internat. Centre Nat. Res. Scient., No. 157, p. 195 (1967).

490. P. Beardmore et al., Trans. Met Soc. AIME, 236 (1):102 (1966).

491. P. L. Clung et al., J. Phys. Chem. Soc., 26 (12):1753 (1965).

492. E. Mooser and W. Pearson, J. Electronics, 1:629 (1956).
493. V. M. Glazov and N. N. Glagoleva, Izv. Akad. Nauk SSSR, Neorgan. Mat., 1 (7):1079 (1965).
494. K. C. Jain and S. Bhan, Trans. Indian Inst. Metals, 19 (3):49 (1966).
495. V. M. Pan et al., in: Metallography, Physical Chemistry, and Physics of Semiconductors [in Russian], Izd. Nauka, Moscow (1967), p. 157.